DOMINATE THE SAT & ACT:

AN UNUSUAL GUIDE FOR UNUSUALLY SUCCESSFUL STUDENTS

MATH EDITION

Written by
BrainStorm CEO and Chief Brainiac Scott Doty

{February 2019}

Good luck, guys.
Go get it.

–Scott

P.S. ~ For even more content and best practices on such topics as time management, productivity, focus, study skills, and more, check out my FREE videos at **thescottdoty.com.**

TABLE OF CONTENTS

"Dominate the SAT & ACT: An Unusual Guide for Unusually Successful Students, Math Edition"
by Scott Doty

www.stormthetest.com
www.thescottdoty.com

© 2019 Scott Doty

Cover by Kristi Redmond

ISBN: 978-1-7330200-0-8

READ ME FIRST!

To my fellow aspiring Brainiac:

WELCOME to my test prep SuperGuide!

I have spent more than 10,000 hours coaching students from all over the world to outstanding performance on game day. This book is the distillation of countless hours of study, research, and experimentation on my part over the course of 15 years as a professional Academic Mentor and Performance Coach.

So: trust the techniques and content offered in this guide. These systems WORK, so long as you work these systems into a daily habit of excellence.

I also want to encourage you to keep perspective. These tests do not– I repeat, do NOT– reflect your value as a human, truly prognosticate your future, or replace the importance of your performance over time in high school. They are simply tests of your knowledge, your reading capacity, and your performance under pressure on a certain date in time. They are a tool, not an enemy.

My recommendation: if you're committed to going to a U.S. university or college, and even if you're leaning towards Test-Optional schools, leave your complaints and excuses at the door and just commit to outstanding effort towards test day. You're holding a guide that will get you to your goals if you do the work. And the knowledge and mindset you develop through the process are yours to keep forever!

I am thrilled and honored you've chosen this book to equip you to <u>STORM</u> the test. You can do it!! Through this process, my desire is that you experience **Dominance on Test Day– & Confidence for Life!**

With enthusiastic nerdiness,

Strategies & Timelines for Junior Year:
Some Organizing Principles

Goal: Our ultimate goal is to achieve admission into a goal college or university, preferably with a merit scholarship. A secondary, but similarly important, goal is to stay sane—and even confident!—throughout the process. How to achieve these ends is, of course, subject to some debate and can furthermore depend heavily on the specific situational needs of the student and his family.

Means to an End: A fabulous SAT &/or ACT score is not the goal; it is the means to the true goal of admission (with money) to a dream school. People ostensibly hire BrainStorm to help them raise their standardized test scores; what they really hire us for is to set them up for success during admission time senior year—and there are myriad ways for us to do so. Some schools are "test optional"... others clearly state that they give more weight to essays and interviews than to test scores... and, with the possible exception of phenomenal athletes, the vast majority of students need to rely more on their transcripts (the confluence of GPA, class rigor, and high school) than on any other admission factor. Could legacy help? Could heritage or state of origin hurt? Could the timing of the application make a difference? In short, the beguiling process of college admission is multifaceted, and does not rely entirely on test scores.

Beliefs based on experience &/or statistical evidence:

- Chances of admission are generally increased significantly (though there are notable exceptions) when one applies "early"—early action, early decision, early in the rolling admission cycle, etc;
- The quality of the senior year experience is typically enhanced when the student has already achieved admission by Christmas, rather than by April;
  - o THEREFORE: Applying early is typically (but not always) advantageous. Writing killer application essay(s), rocking the interviews (when offered) and campus visits, gathering the recommendations, preparing the CVs, confirming the list of goal schools, and filling out the applications themselves: this is the Herculean task of the inchoate senior. To excel at all aspects of this process without falling behind in rigorous senior-year classes (to say nothing of keeping up with extra-curriculars), it is advisable that students avoid taking standardized tests in the fall. ("Spread thin," I believe would be the appropriate idiom.) The typical loss of academic momentum accrued throughout the summer also figures in this equation. As such, fall of senior year is our back-up option as needed—but it is not part of the ideal timeline.

- For tremendous athletes who are being recruited, the earlier the SAT/ACT test scores become available, the better. Example: My student, Rae C, committed to Holy Cross this spring because, by prepping early & often with me during junior year, she had already achieved HC's required minimum

score by March. Better scores earlier in the year more easily allow college coaches to make commitments to student-athletes.

- o THEREFORE: The earlier we get a solid, usable score on the board for a student-athlete, the better.

- Because final junior year grades are so vital to the applicant's ability to earn admission, June of junior year should be devoted to finishing the fourth marking period strong & to acing final exams. May is the time for AP exams for elite students.

- o THEREFORE: June SAT/ACT tests are back-up options, not ideals. For AP-level students, May &/or June should be used for SAT Subject Tests. For non-AP students, May is the goal month for the last SAT sitting. For both sets of students, April is the goal month for the last ACT sitting. {Now that each test is offered in the summer, that's a decent back-up option.}

- Parkinson's Law states that "Work expands so as to fill the time available for its completion." What to do when a client hires us in September of junior year, or even earlier? BrainStorm's job is to motivate & organize the student, and to give him tools that will make his individual study time more prosperous. What happens when I tell my callow 16-year-old friend that our first "real" test will not take place until 4-5 months down the road? He loses motivation. He'll take the same amount of work he could have achieved over 8 weeks and spread it over 18. Committing him to an earlier deadline—a test that matters that's right around the corner—imbues the relationship with a sense of urgency & focus, two of the student's most useful tools.

- The PSAT is useful only for elite students who have a shot at the National Merit Scholarship. Other students know the results are not crucial, so many do not put forth honest effort, while those who do will not receive feedback for 2 whole months—a monstrous chasm of time during a fraught junior year. Are the results even accurate as a diagnostic tool eight weeks after the fact? Are the results useful, period? The PSAT does not test the essay, the student's ability to perform under legitimate pressure, or the student's ability to focus and perform over a grueling 4-hour test. The ACT PLAN test is closer to useful, but similarly underperforms.

- o THEREFORE: We use full-length SAT & ACT simulations as diagnostics. We ask students to take them early & often. And when the timeline is significantly expansive, we create short-term deadlines that impel effort—even if it means signing up for fall sittings that we may very well push off when the date arrives. At least the specter of an imminent test gets us out of the gate with forceful energy! Further, sometimes there is no substitute for the genuine article—taking the "real deal" SAT/ACT, with stakes and all, helps certain students get their head around the experience, and makes their subsequent work more purposeful. Finally, most schools now offer "Score Choice"—yet another reason to feel comfortable experimenting with fall/winter sittings as a junior.

NOVEMBER-APRIL
OF JUNIOR YEAR FOR TYPICAL STUDENTS

OCTOBER-MARCH
OF JUNIOR YEAR FOR ELITE STUDENTS WHO ALSO PARTICIPATE IN A DEMANDING SPRING SPORT

MAY & JUNE ARE FOR SCHOOL WORK
AND POSSIBLY APs AND/OR SAT SUBJECT TESTS

JULY-OCTOBER
IS FOR SCHOOL VISITS & APPLICATIONS (AND FOR ADDITIONAL SAT/ACT/SAT2 SITTINGS AS NEEDED)

Possible Objections:

- Doesn't the student need a full year's worth of Algebra 2 to be able to tackle the SAT/ACT Math?

  NO. The ACT Math features about 15-18 Alg2 questions on just a handful of topics (basic trig, logs, imaginaries, etc), and the SAT Math, as a percentage of total questions, features Alg2 about equally. The number of Alg2 topics required for both tests is miniscule relative to the raft of Alg1, Data, and Geometry topics required, and can be taught quickly via private tutoring rather than waiting for an eventual (and hopefully, but not necessarily, competent) school lesson among a group of 20 kids. And don't forget about KhanAcademy.org! Many of our students have gotten way ahead in school by plowing through KA's math modules (for free). Furthermore, elite students—who need Alg2 to earn elite scores—take the course freshman or sophomore year; lower-caliber students don't need that minority of test questions focused on Alg2 to hit their (lower) goal scores. **Bottom line:** Having done an entire school year of Algebra 2 is preferable but not necessary; there are legitimate work-arounds.

- Doesn't waiting until later in junior year help the student to increase his vocabulary?

  There is no vocabulary on the ACT, so the point is irrelevant with regard to that test. For the SAT, yes: more months in school implies more vocabulary. Or at least, the extra time increases the number of words our student has memorized, regurgitated for a quiz, and forgotten about along the way. But fair enough: vocabulary should be better in March than it is in November. Studies show, however, that teenagers' ability to recall memorized information like word definitions diminishes during periods of high stress (such as that experienced in the last eight weeks of junior year) and following an academically barren season (like summer break). In our experience, vocabulary utility & recall hits its relative peak in April of junior year. Finally, even the SAT focuses far less on vocabulary than it used to. **Bottom line:** vocabulary is not an important enough factor in SAT/ACT performance to alter the timeline recommended above.

- What if the student is considering applying to schools that don't accept Score Choice?
  The vast majority of schools do offer Score Choice for the SAT. Just about all schools offer it for the ACT. Those schools that do not accept SAT's SC—everyone from Colgate to CUNY, Louisiana State to McGill—do give us pause when it comes to scheduling a first sitting. In this case, we weigh the potential benefits of an early sitting—those articulated above—with the potential costs, and make a call based on how quickly the student is making progress.

- Exceptions: What about kids who need first semester grades from senior year to increase their chances of school admission? What about kids trying to avoid ED apps so as to have a better shot at earning merit scholarship monies? What about kids applying to schools at which the early acceptance rate is lower than that of regular decision? What about kids who excel in a winter sport and would like to increase their chances of getting recruited? Etc.

  Indeed, there are many cases in which waiting until regular application deadlines (typically January) is advantageous for the student. For these students, taking the SAT/ACT in fall of senior year (in an effort to raise those junior year scores just that extra notch) makes sense. However, these are the exceptions to the rule. The rule is this: avoid standardized during senior year. There is already an exasperating litany of tasks to execute during this time, and a mountain of attendant stress. View the SAT & ACT for what they are: two of the many ways to earn admission to & merit monies from a tremendous undergraduate institution. Knock them out in junior year and get them off your plate of to-do's. Remember that it's wiser to plan this way, and to use fall of senior year as a back-up, than to plan to earn your breakthrough score during fall of senior year—and then have no viable back-up option, time-wise, if you don't succeed.

YOU GOT THIS!!
MAKE THE PLAN, THEN WORK THE PLAN,
& YOU WILL EXPERIENCE DOMINANCE ON
TEST DAY & CONFIDENCE FOR LIFE!! -Scott

RULES OF GAME DAY: How to STORM the Test
& Perform at Your Best When the Stakes Are High

I.   One Focus at a Time

Success on test day includes many, many factors. We will outline many of the logistically clever things to do to best ensure a high probability of success, but let's begin with a broader, more overarching philosophical point that runs counter to popular understanding: the human mind cannot truly multitask. The brain is a sequential processor—if we wish it to accomplish 7 things, we must allow it first to focus on the initial task until that job has been performed; then we can push it towards the second task, etc. Unless we wish for a maelstrom of shoddy work, we cannot ask the brain to perform all seven tasks—or even two of these tasks!—at once. There is much literature on the subject in support of this truth.[1]

Take, as an example, a much-maligned pastime: texting while driving. Are we really, truly able to drive a car and text a friend simultaneously—or, more probable, are we able to focus on driving, then briefly focus on texting (neglecting the road for a few seconds), and then return to driving? I hope we all agree intuitively—if not by experience—that the latter is true. This is why

---

[1] See the "Further Resources" section at bottom for a starting point on the research.

texting while driving is so dangerous! We really cannot do both with high quality at the same time.

This important truth—that the mind must focus on one thing at a time if it is to perform well—informs and edifies my main test-taking philosophy: test success relies on creating and then maintaining the most beneficial focal point for the mind. Remember, there can only be one!

~ONE focal point for your ONE mind at any ONE time~

Unfortunately, most people choose the wrong one. In fact, most people do not consciously "choose" one at all. Many people give conscious thought only to the academic material that will be present on the test; however, few give even a modicum of effort to emotional and psychological readiness, which are typically LARGER contributors to success than is academic knowledge.[2] As such, we see intelligent, diligent students nettled, flummoxed, and finally undone by test day anxieties (and other focus miscues). Our goal here is to DESTROY the entrenched viewpoint that tests are a purely academic exercise and to PROPEL students into a more holistic, pragmatic, and effective study regimen that includes conscientious endeavor in non-academic fields. Beyond these, our ultimate goal is to create life opportunities for students by empowering them to STORM THE TEST!

II. Prevailing Mentality Faux Pas

Emotional focal points, when reinforced over time, become what I call "prevailing mentalities." Is your prevailing mentality in life generally one of

---

[2] I should think this is intuitive—how are you to perform well when you're stressed out of your mind or terribly distracted?—but for many people it is not. See "Further Resources."

"The world is a safe place; I feel secure and happy"? Or is it closer to "I am unworthy of respect; I am not intelligent", or perhaps "I'm afraid of taking risks"? Whatever our typical mentality is in life, we can work—through conscientious endeavor—to achieve a prevailing mentality on test day that is most advantageous.

Before unveiling to you what I believe to be the most beneficial prevailing mentality—the one that unlocks doors to incredible personal success—I would first like to point out typical "prevailing mentality faux pas." Typically, people choose (largely unconsciously) a prevailing mentality the day of a test that is insidious and undermining. The most common of these are Quality, Clock, Stakes, and History.

A **Quality focus** sounds promising, but it is not. People whose emotional commitment on test day is toward quality often struggle: they over-commit to any one question and get lost in it at the expense of the bigger picture, and they often run out of time. They also can become extremely frustrated when they struggle with as few as two or three questions in a row, as this blatantly undermines their main goal of getting every question fantastically, flawlessly correct. People in this category make great physicists or engineers because of their commitment to quality and precision—but they are not gifted test takers.

A **Clock focus** implies that the test taker is obsessed with the fear of running out of time, often to the extent that he or she is unable to truly engage with the academic material because of a constant time preoccupation. Running out of time is the cardinal sin of test taking, this student thinks. This person may indeed hit the goal of finishing on time, but at what cost? Typically, at the very high cost of sacrificed quality. On certain standardized tests like the SAT, finishing on time is far from the main strategic goal—but this strategic reality is lost on a clock-focused test taker.

A Stakes focus is an equally detrimental prevailing mentality, for it keeps the test taker focused outside of the present moment. It is the constant, badgering inner voice that reminds the test taker that the current test has HUGE future implications. The imposing specter of the test's implications completely consumes the test taker's emotional energy and conscious focus. Instead of looking forward to the test with joy and optimism (which, as I explain later, can be done!), he or she dreads the test, fears it—and makes it bigger than it really is. These test takers are typically quite nervous, fidgety, and scatterbrained—only to become lucid minutes after the test booklet has been officially handed to the proctor.

A History focus is tantamount to a Stakes focus in that it robs the test taker of relaxed, here-and-now concentration. In this mentality, the test taker focuses on her history of tests, which is (in her mind, anyway) somewhat deplorable. She owns and has internalized the self-imposed aspersion "I am terrible at tests." She looks backwards: yesterday's failures prove conclusively that tomorrow is just another opportunity to fail. This self-denigrating attitude de-motivates the student during the weeks & months of test prep (Why should I even try? It's fruitless) and becomes a self-fulfilling prophecy.

All of these mentalities are destructive, and answer the oft-asked "If I am generally a very intelligent person, why do I struggle so much on tests?" There are other destructive prevailing mentalities, of course—fear of others' judgments of results, indolence and avoidance of self-challenge, etc—but I find the four mentioned above to be most prevalent and pernicious.

III. Let this Mentality Prevail! <u>Optimistic Momentum</u>

The good news: we CAN consciously choose our prevailing mentality for the day. We can coax it out of ourselves, and we can maintain it once we have established it. If we execute this process enough times, it becomes an ingrained habit that is unstoppable in the most wonderful way. The prevailing mentality we want to engender on test day is what we will call OPTIMISTIC MOMENTUM.

Optimistic momentum (OM) implies the following: a healthy, balanced understanding of the stakes of the test; a confidence that everything will work for the good in the big picture, regardless of results; a heightened sense of emotional energy and optimism; a razor-sharp, disciplined strategy based on both exam structure and personal rhythm; a commitment to the quality of the test day process, rather than test day numerical results; and, above all, an enlarged sense of joy for the grand, wonderful opportunity to take this test—because it is going to be so much freakin' FUN! This last point, of course, sounds especially laughable for most people, but it is achievable—and when people train themselves to view a challenge as fun, keeping in mind that the stakes are not as big as they think anyway, true breakthrough scores are feasible. Truly, people should look forward to test day, rather than loathe it—this test is one of many, and life goes on afterwards.

*~Being confident, like being nervous, is a decision.
Why not prepare well and then go in and STORM it?~*

During the test, a person employing optimistic momentum sticks to strategy, technique, and pacing without over-thinking any of these; he cares more about maintaining his unflappable mentality than getting any particular question correct; he is resilient, recovering from temporary setbacks by breathing, smiling, retaining perspective, and giving himself a pep talk (That was not ideal, but no worries. I'm enjoying myself; let's keep moving.). When optimistic momentum is properly executed, quality and time management are naturally taken care of—there is no need to focus on them. Focus is instead placed on enjoying the test, being emotionally present and self-confident, and allowing these good things to follow naturally.

IV. **Before test day**

From this strong place of aiming to achieve optimistic momentum—the kind of good vibes that propels one through a test with vim—we create logistical steps:

- First, if we have a number of days or even weeks or months to prepare for Test X, we begin by doing a personal inventory: What is my typical prevailing mentality on test day? Become self-aware and begin taking pragmatic steps to arrest the persistence of unwanted habits: start a journal, get an accountability partner, utilize symbolic measures ("I'm using this new pencil and wearing these new shoes to symbolize that this is a new day!"), etc.
- Second, learn the invaluable skill of giving yourself pep talks. Seriously.
- Third, attack the academic material in earnest and learn the strategic pillars of this particular test; inculcate yourself, gaining

15

deep comfort with the material & strategy so that on test day you need not worry about recalling them. These things should just "be there." Pre-emptive excuses not to study aggressively are unacceptable; they express only that the student retains a destructive prevailing mentality of which he must be relieved. (NOTE TO THE INSTRUCTOR: This task of empowering students to disavow undermining thoughts is far more difficult, but far more enriching to the student, than simply teaching academic material. Challenge yourself to focus energies on embodying & teaching OM—become a motivation <u>master</u>—and trust that the academic material will get itself across along the way.)

- o Fourth, take as many full-length simulations as possible, focusing energies on the skill of creating and utilizing OM rather than on numerical results. Learn to trust yourself: the results will follow.

The night before a test, should we cram? Absolutely not! This only serves to feed one of the other, destructive mentalities—often that of Stakes. Do the work properly during the weeks and months prior to test day, and then use the night prior to establish OM. For me, this means watching a goofy movie (I use "Dumb and Dumber") with a good friend, and laughing a lot. I give my students a lot of latitude on this front—so long as they're getting into a positive mentality, I am pleased.

Going to sleep much earlier than usual is discouraged—if the body has not been physiologically trained to sleep at 8pm, why get in bed at that hour? Chances are good you will simply lie there, growing frustrated and anxious. Instead, go to sleep perhaps just a bit earlier than usual, confident that you will fall sleep quickly.

V. Test day

We will assume for this discussion that the test begins in the early morning (8am). Most of these principles help whether the test takes place at this hour or not.

 PRE-TEST

WAKE UP: No later than 6am. 5:45am is ideal. The typical adult brain requires 70 minutes to be fully alert; the typical teenage brain requires 90-100. People think they should sleep as much as possible before the test to ensure they are well rested—but this approach is counter-productive. Give the mind time to wake up, so you do not bomb the very first section of the test—which, as we can imagine, sets a terrible emotional tone for the rest of the day. Note that the most important nights for deep, curative sleep are the Tuesday, Wednesday, and Thursday nights leading up to a Saturday test.

WORK OUT: Do 20 minutes of physical exercise. This does not mean heavy lifting, but it does mean stretching, jogging, push-ups, yoga, etc. The action has two crucial benefits: it moves oxygen to the brain, which it needs to properly wake up, and it raises the levels of endorphins—the chemical that makes us "happy." I often encourage people to listen to uplifting, powerful songs while they do this wake-up workout.

BREAKFAST: Stick roughly to your norm. Sure, it's not ideal to drink coffee or eat sugary treats—but if your body is accustomed to them, it may have problems getting out of the gate without them. The ideal is to start on a healthy regimen a week or two before the test—protein & quality fats (via eggs, avocado, nuts, fish) and high-vitamin fruit (bananas, berries) are good.

The key is focus on hydration and low glycemic-index foods that give a long, smooth burn of energy instead of a quick peak and crash.

SHOWER: If the person showers that morning, he or she should not let the shower be uber-hot and long... too easy to fall back into sluggish mode. We are going to dominate today! Life is good! We take COLD showers, or at least ensure that we finish our typical warm shower with at least 4 minutes of awesomely cold water. The results are incredible~ trust me!

CLOTHING: This point is non-negotiable: you MUST wear smart clothes. No sweatpants or pajamas, which your brain has been trained to associate with tiredness and sloth. Instead, we do what professional athletes do when they travel between cities: we dress sharp, because it sharpens the mind. Slacks, button-downs, nice shoes, even hair gel and cologne/perfume—these things tell us to be sharp, to be aggressive, to be professional. We are here to kick butt and take names. We also wear/bring layers so we can adjust to unforeseen room climates.

RELAX: Meditate, pray, do some light problems, give yourself some pep talks. Smile. Tackle a puppy. It's time to keep up those good vibes you established last night.

WHAT TO BRING: First, the two essentials are your printed ticket and a photo I.D. To those, add academic extras (extra calculator, batteries, pencils, etc); a stopwatch if helpful (make sure it's on silent mode); and a "bag of tricks"—snacks, water, mouthwash, Altoids or other potent breath mints, headache medicine (I personally use Excedrin because it also has a dash of caffeine), a laminated picture of Sheldon Cooper from "Big Bang Theory" for inspiration, etc. And leave the phone at home. Don't even bring it inside & turn it off—just leave it behind.

✦ DURING THE TEST

If pre-test is properly executed, the test taker should be awake, calm, and optimistic. He or she will probably notice that the other test takers are far less so, and this will only serve to augment our test taker's confidence. We should have no problem jamming through the first two hours of the test. We will balance intense focus with clean, controlled rest throughout the rest of the day, and will aim for a consistent performance: no huge peaks or troughs. We want to hit our groove from before the first minute until after the test has finished—at which point we can dance salsa in the hallways or crash on our friend's couch for a massive nap. Whatever.

Below is a diagram depicting the (unfortunate) typical test day experience for most students. It shows what happens when the foregoing test day rules are not followed scrupulously—and the natural result of corrosive prevailing mentalities. I have dubbed the diagram, because of its form, "Elephant Under a Bridge."

ELEPHANT UNDER A BRIDGE
Diagram of a Student's Typical Test Day Performance

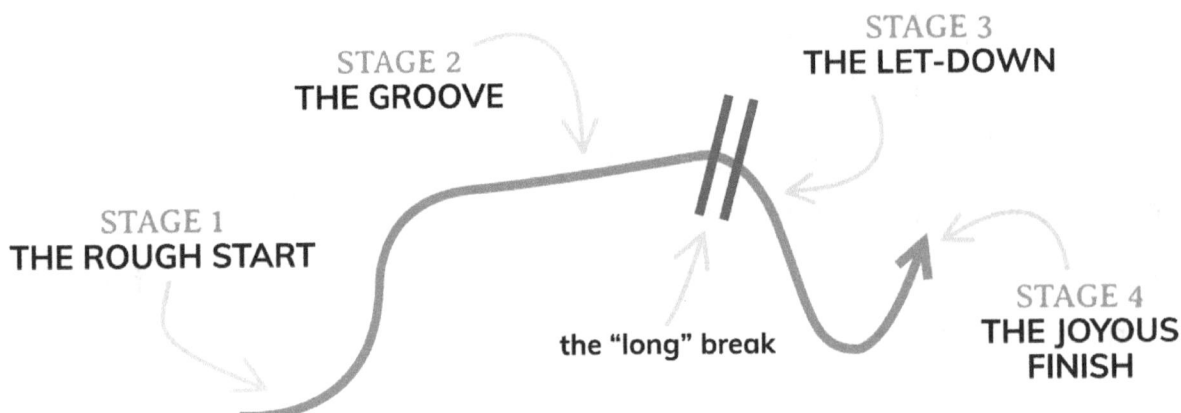

STAGE 2
THE GROOVE

STAGE 3
THE LET-DOWN

STAGE 1
THE ROUGH START

the "long" break

STAGE 4
THE JOYOUS FINISH

STAGE 1: THE ROUGH START
Students often make one of the following mistakes: staying up too late or going to bed too early, cramming, skipping breakfast, waking up just in time to run out the door, etc. Students therefore typically begin the test in a foggy, stressed, or nervous mindset that lasts for a solid 45 minutes. These and other missteps play a key role in preventing a student from kicking off the test at the optimal level.

STAGE 2: THE GROOVE
The student is now fully alert and focused and is hitting a bit of a stride. This lasts for a solid 75-100 minutes, during which confidence builds. Soon enough, however, the student finds himself experiencing...

STAGE 3: THE LET-DOWN
Stress, lack of adequate morning-of preparation, and improper emotional self-management cause many students to experience a crash in energy & focus when they're about one-half to two-thirds of the way through the test. This is precipitated by the "long" break of 5-10 minutes afforded to the students at that time. Students suffer a loss of momentum and vigor, and sense an ebb in their emotional commitment to succeed.

STAGE 4: THE JOYOUS FINISH
After the let-down, the student suddenly recognizes he is almost done! He's in the home stretch! Adrenaline and excitement return for a solid finish.

Some particularly malevolent tests (SAT & ACT included) exacerbate the issue by intentionally placing the most challenging sections at the juncture where students might struggle anyway with fatigue, nerves, and lack of focus. Notice, for example, that the SAT's first section is the painfully long Critical Reading section and its hardest section, the first math test, is placed third in order. The ACT's hardest section, the crazy-fast Critical Reading test, is also intentionally placed third in order. Those test makers are evil!

Of course, the test is simply doing its job: separating the strong test takers from the weak test takers. Notice they are NOT separating the more intelligent from the less intelligent! These exams either test for aptitude (read: intellectual creativity and poise, à la SAT) or for achievement (read: knowledge base and speed of recall, à la ACT)—they cannot test for pure intelligence. So stop yammering about the disjoint between your intelligence and your scores! Realize that the test is investigating something OTHER THAN RAW INTELLIGENCE—and now go work on that thing.

* * * *

✧ OTHER TEST DAY POINTERS

We take breaks seriously—especially the one in the middle of the test. We need to avoid hitting a physical or emotional energy wall, so we use the break to gear back up. How? By reestablishing some pivot points we hit that morning: by moving oxygen in our systems (jumping jacks in the hall, push-ups in the bathroom), by eating/drinking a bit, by giving ourselves a pep talk and high-fiving a buddy ("Isn't this fun, bro?"), and by delving into our bag of tricks (taking an Excedrin, popping a handful of Altoids, kissing Sheldon, etc). We are doing well!! We are enjoying ourselves! Section 2 was not our best, but it's all good—we will finish strong and let the chips fall where they may. We are engaged in the moment—not over-thinking the stakes, not looking forward to being done with this wretched thing... we are right here, taking care of business. Overarchingly, we know our self-worth and professional future do not hang on the results of this one test, so we're willing to remain emotionally open to focusing until the last bell—even if a certain section along the way has been disappointing.

If we can start strong and avoid hitting the wall around the two-thirds mark, we have got it made. If we allow our general enthusiasm and optimism surrounding the experience to suffuse our mentality, we will not agonize over individual questions—none of which makes or breaks the overall score—but rather briskly, confidently rock from one question to the next and keep the vibes positive. In short, if we manage to create optimistic momentum the night before the test, and carry that momentum through the test day experience itself, then we will have truly given ourselves the best possible chance at succeeding.

What if my phone goes off during the test? Don't worry. You left it at home. What if the wall clock is broken? No worries: you brought a watch on silent mode. You also should ask the proctor for time updates throughout the section.

What if a fellow student is annoying me? Ask the person directly to stop smacking his gum or jittering his leg; ask the proctor to intervene if need be; ask to move seats; bring earplugs.

What if I look stupid wearing smart clothes? If wearing a certain article of clothing for 5 hours helps you get a massive score jump, is it not worth it??

What if I lose my Sheldon Cooper picture? Then today is not your day, my friend.

VI. SUMMARY

We have made the following points:

o   The human mind can only focus on one thing at a time

o   The mind's emotional focal point, when reaffirmed over time, becomes a prevailing mentality that frames our intellectual process

o   Despite ingrained habits, we have the power to consciously elect and embody the prevailing mentality that will most benefit us

o   For tests of all kinds—from typical school tests to major standardized tests—the ideal prevailing mentality is Optimistic Momentum (OM)

o   When combined with rigorous academic and strategic preparation, OM empowers students of all learning styles and calibers to achieve breakthrough test performance

o   Detail-oriented logistical measures during the weeks leading up to test day and during test day itself provide a disciplined structure that further enhances performance

Ultimately, academic evaluations test far more than our ability to retain and regurgitate a host of data. Like any other challenge, they give us the opportunity to hone and employ life skills that far transcend in importance the immediacy of numerical results. If we focus on the transcendent qualities discussed here—relaxed concentration, ferocious optimism, the joy of meeting a challenge with our best effort—we become free to gain the external

victories we also seek (huge score gains, school acceptance letters, etc) without trying so hard to get them. As W. Timothy Gallwey wrote, "It is the moment-by-moment effort to let go and to stay centered in the here-and-now action which offers the real winning and losing, and this game never ends."

FURTHER RESOURCES:

On the brain's inability to multitask:

http://www.startupnation.com/series/127/9293/multitasking.htm

http://brainrules.blogspot.com/2008/03/brain-cannot-multitask_16.html

http://www.youtube.com/watch?v=BpD3PxrgICU

http://www.npr.org/templates/story/story.php?storyId=95256794

On the effect of stress on focus and test-taking abilities:

http://www.education.com/reference/article/Ref_Helping_Overcome/

http://kops.ub.uni-konstanz.de/bitstream/handle/urn:nbn:de:bsz:352-opus-769 62/Goetz_et_al_LEAIND.pdf?sequence=1

(Or search "Thomas Goetz, et al, 'Emotional experience during test taking', 2007")

http://www.amsciepub.com/doi/abs/10.2466/pr0.1967.20.3.975

http://academicanxiety.org/wp-content/uploads/2011/08/cassady-johnson-20 04.pdf

A fantastic book on the general topic: The Inner Game of Tennis by W. Timothy Gallwey

SAT/ACT I.Q. TEST

by Scott Doty

1. What does "Score Choice" mean?

2. What is the meaning of "Super Score" and on which test is it used?

3. On what websites does one sign up for the SAT and the ACT?

4. What is the difference between the math sections of the two tests?

5. Which test does better to earn college admission?

6. Which test has been around for more years? Historically, where has the ACT been accepted, and where has the SAT been accepted?

7. In what order are the sections in the SAT? what about the ACT?

8. Which test is easier?

9. Which test is shorter?

10. What's the difference in blanks strategy between the two tests?

11. How many times each year is the SAT offered? How about the ACT?

12. What is the difference between Early Action and Early Decision college applications?

13. What are "rolling admission" applications all about?

SAT/ACT I.Q. TEST ANSWERS

1. Score Choice is the ability to withhold test results from a certain SAT/ACT date from schools to which you are applying. The vast majority of schools accept it~ meaning that at the vast majority of schools, you only need to send in the results from dates you choose.

2. Super Score is what's created when you combine sub-section scores from different dates. For example, in a Super Score you can take your math score from one date and pair it with your verbal score from another date, and there you have your best overall "Super" score for consideration. The minority of U.S. universities and colleges accept SuperScore. More accept it for SAT than for ACT.

3. You can sign up for the SAT on www.collegeboard.org and for the ACT on www.act.org.

4. The math of the SAT is in two sections, one of which restricts the use of a calculator. In each of those two sections there are multiple choice questions and grid-in responses. The test focuses on fewer math topics and requires that few formulas be memorized, but asks questions that are more wordy/challenging. The math of the ACT, conversely, is concentrated in one section and is entirely multiple choice and calculator-friendly. The test gives relatively straightforward problems, but the student is expected to know a lot of content and a lot of formulas.

5. Neither! They are equivalent in the eyes of admissions counselors.

6. The SAT pre-dates the ACT by a few decades. Historically, the SAT was more widely accepted by schools in California and the Northeast, whereas the ACT was more widely accepted in the Midwest and Deep South.

7. The SAT's order is Critical Reading, Writing & Language, Math One, Math Two. The ACT's is English, Math, Critical Reading, Science.

8. This answer depends on the student. Most people find the ACT easier in terms of the difficulty of its questions but harder in its pacing. For those students who are fast readers and/or have IEPs (and therefore qualify for extended time), the ACT is the recommended option. For people who are quite slow readers, especially those deeply intimidated by the ACT Science, the SAT is recommended.

9. By just about 15 minutes when including the essay, the ACT is shorter.

10. There is no difference in blanks strategy between the tests. In both cases, no points are lost for wrong answers, so students are encouraged to leave ZERO blanks.

11. The SAT is currently offered in March, May, June, August, October, November, & December (7 times per year). The ACT is offered February, April, June, July, September, October, & December (7 times per year).

12. Both typically have application deadlines around November 1. Early Decision is a binding contract: if a student applies ED to a certain school, she may NOT apply ED to any other school, and if she is accepted, she *must* attend. Early Action, on the other hand, is NOT a binding contract, so students are permitted to as many Early Action schools as they like.

13. Rolling applications are simply a system by which a college or university allows people to apply at any time after a launch date. There is no deadline for applying, though the sooner the student applies, the better.

ACT°–SAT° Concordance: A Tool for Comparing Scores

The ACT° college readiness assessment and SAT° are different tests that measure similar but distinct constructs. The ACT measures achievement related to high school curricula, while the SAT measures general verbal and quantitative reasoning.

ACT and the College Board (producers of the SAT) have completed a concordance study that is designed to examine the relationship between two scores on the ACT and SAT. These concordance tables do not equate scores, but rather provide a tool for finding comparable scores.

You can also find the concordance tables and guidelines for proper use on our website at **www.act.org/aap/concordance**.

ACT Composite Score	SAT Score Critical Reading + Math (Single Score)	SAT Score Critical Reading + Math (Score Range)	ACT Score Combined English/Writing	SAT Score Writing (Single Score)	SAT Score Writing (Score Range)
36	1600	1600	36	800	800
35	1560	1540–1590	35	800	800
34	1510	1490–1530	34	770	770–790
33	1460	1440–1480	33	740	730–760
32	1420	1400–1430	32	720	710–720
31	1380	1360–1390	31	690	690–700
30	1340	1330–1350	30	670	660–680
29	1300	1290–1320	29	650	640–650
28	1260	1250–1280	28	630	620–630
27	1220	1210–1240	27	610	610
26	1190	1170–1200	26	590	590–600
25	1150	1130–1160	25	570	570–580
24	1110	1090–1120	24	550	550–560
23	1070	1050–1080	23	530	530–540
22	1030	1020–1040	22	510	510–520
21	990	980–1010	21	490	480–500
20	950	940–970	20	470	470
19	910	900–930	19	450	450–460
18	870	860–890	18	430	430–440
17	830	820–850	17	420	410–420
16	790	770–810	16	400	390–400
15	740	720–760	15	380	380
14	690	670–710	14	360	360–370
13	640	620–660	13	340	340–350
12	590	560–610	12	330	320–330
11	530	510–550	11	310	300–310

The**ACT**°

www.act.org/concordance

TABLE 1

Explanation of Procedures Used to Obtain Scale Scores from Raw Scores

On each of the four multiple-choice tests on which you marked any responses, the total number of correct responses yields a raw score. Use the table below to convert your raw scores to scale scores. For each test, locate and circle your raw score or the range of raw scores that includes it in the table below. Then, read across to either outside column of the table and circle the scale score that corresponds to that raw score. As you determine your scale scores, enter them in the blanks provided on the right. The highest possible scale score for each test is 36. The lowest possible scale score for any test on which you marked any responses is 1.

Next, compute the Composite score by averaging the four scale scores. To do this, add your four scale scores and divide the sum by 4. If the resulting number ends in a fraction, round it to the nearest whole number. (Round down any fraction less than one-half; round up any fraction that is one-half or more.) Enter this number in the blank. This is your Composite score. The highest possible Composite score is 36. The lowest possible Composite score is 1.

ACT Test 67C	Your Scale Score
English	_____
Mathematics	_____
Reading	_____
Science	_____
Sum of scores	_____
Composite score (sum ÷ 4)	_____

NOTE: If you left a test completely blank and marked no items, do not list a scale score for that test. If any test was completely blank, do not calculate a Composite score.

Scale Score	Test 1 English	Test 2 Mathematics	Test 3 Reading	Test 4 Science	Scale Score
36	75	59-60	40	40	36
35	73-74	57-58	39	39	35
34	71-72	55-56	38	38	34
33	70	54	—	37	33
32	69	53	37	—	32
31	68	52	36	36	31
30	67	50-51	35	35	30
29	66	49	34	34	29
28	64-65	47-48	33	33	28
27	62-63	45-46	32	31-32	27
26	60-61	43-44	31	30	26
25	58-59	41-42	30	28-29	25
24	56-57	38-40	29	26-27	24
23	53-55	36-37	27-28	24-25	23
22	51-52	34-35	26	23	22
21	48-50	33	25	21-22	21
20	45-47	31-32	23-24	19-20	20
19	42-44	29-30	22	17-18	19
18	40-41	27-28	20-21	16	18
17	38-39	24-26	19	14-15	17
16	35-37	19-23	18	13	16
15	33-34	15-18	16-17	12	15
14	30-32	12-14	14-15	11	14
13	29	10-11	13	10	13
12	27-28	8-9	11-12	9	12
11	25-26	6-7	9-10	8	11
10	23-24	5	8	7	10
9	20-22	4	7	6	9
8	17-19	—	6	5	8
7	14-16	3	5	4	7
6	11-13	—	4	3	6
5	9-10	2	3	—	5
4	6-8	—	—	2	4
3	5	1	2	1	3
2	3-4	—	1	—	2
1	0-2	0	0	0	1

The **ACT** Student Report

ANN C TAYLOR (ACT ID: -54116290)
WHEAT RIDGE SR HIGH SCHOOL (061-450) | APR 2016 NATIONAL

Composite Score | **21** | U.S. Rank 56% | State Rank 58%

Test Results

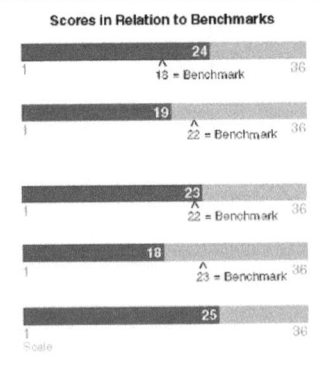

Test Results	Score	U.S. Rank
English	24	74%
Usage/Mechanics	12	72%
Rhetorical Skills	12	71%
Mathematics	19	47%
Pre-Algebra/Elem. Algebra	11	57%
Algebra/Coord. Geometry	10	51%
Plane Geometry/Trig.	09	39%
Reading	23	66%
Social Studies/Sciences	12	67%
Arts/Literature	11	58%
Science	18	32%
Writing	25	79%
Ideas and Analysis	10	
Development and Support	08	
Organization	07	
Language Use and Conventions	08	

Scores in Relation to Benchmarks:
- English: 24 (18 = Benchmark)
- Mathematics: 19 (22 = Benchmark)
- Reading: 23 (22 = Benchmark)
- Science: 18 (23 = Benchmark)
- Writing: 25

	Score	U.S. Rank
ELA	24	70%
STEM	19	40%

Understanding Complex Texts
Proficient
(Below Proficient — Proficient — Above Proficient)

Progress Toward Career Readiness
You are making progress toward a Gold level on the ACT NCRC.
(Bronze, Silver, Gold)

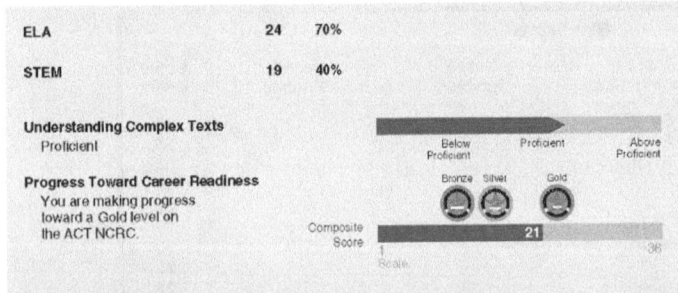

Composite Score: 21

Composite and Subscores: ACT test scores and the Composite score range from 1 to 36; subscores range from 1 to 18. Your Composite score is the average of your scores on the four subject tests. Subscores do not necessarily add up to your score for a subject test.

ACT College Readiness Benchmarks: If your scores are at or above the ACT benchmark scores, you will likely be ready for first-year college courses.

U.S. Rank and State Rank: Your ranks tell you the approximate percentages of recent high school graduates in the U.S. and your state who took the ACT and received scores that are the same as or lower than yours.

Interpreting Your Scores: Test scores are not precise measures of your educational development. ACT scores reported are the midpoint of a score range that represents your educational development at the time you took the ACT. For example, the score range is plus or minus one point for the Composite score. You will find more information about interpreting your scores in the *Using Your ACT Results* booklet provided with this report and at **www.actstudent.org**.

Writing: The score ranges from 1 to 36. Writing domain scores range from 2 to 12. Domain scores do not necessarily add up to your score for the Writing test.

English Language Arts (ELA): An average of your English, Reading, and Writing scores. The score ranges from 1 to 36.

Science, Technology, Engineering, and Mathematics (STEM): An average of your Math and Science scores. The score ranges from 1 to 36.

Understanding Complex Texts: Measures level of proficiency on a subset of items in the Reading test assessing the ability to identify the central meaning and purposes for a range of increasingly complex texts.

Progress Toward Career Readiness: Based on your ACT Composite score, Progress Toward Career Readiness is an indicator of your potential level of achievement on the ACT National Career Readiness Certificate™ (ACT NCRC®). The ACT NCRC is an assessment-based credential that certifies skills critical to your future education and career success.

Learn how ACT NCRC performance relates to job skill requirements at **www.act.org/workkeys/briefs/files/NCRCRequirements.pdf**.

This information is not to be considered a substitute for actual performance on the ACT NCRC.

Your College Reports

At your direction, your scores from this test date are being reported to the colleges shown below. College planning information is provided for the first four choices you listed when you registered or tested. Check with colleges for recent changes in information. Note: Your GPA was calculated from the grades you reported.

College Name (Code)	Profile of Enrolled 1st-Year Students				Approximate Annual Tuition and Fees		Percentage of 1st-Year Students Receiving Financial Aid	
	ACT Composite Score	High School Class Rank	High School GPA	Preferred Program of Study Availability	In-state	Out-of-state	Need-based	Merit-based
UNIVERSITY OF OMEGA (9521) OMEGA, CO WWW.UNIVERSITYOFOMEGA.EDU	MIDDLE 50% BETWEEN 18–24	MAJORITY IN TOP 50%	2.76	4-YR DEGREE	$5,600	$12,000	67%	20%
ALPHA UNIVERSITY (9059) UNIVERSITY CENTER, IA WWW.ALPHA.EDU	MIDDLE 50% BETWEEN 21–26	MAJORITY IN TOP 25%	3.12	4-YR DEGREE	$9,000*	$15,000*	85%	27%
BETA COMMUNITY COLLEGE (8866) CLARKSTON, CO WWW.BETACC.EDU	MIDDLE 50% BETWEEN 16–21	MAJORITY IN TOP 75%	2.49	PROGRAM AVAILABLE	$4,000	$4,000	58%	18%
MAGNA COLLEGE (8905) PLAINVIEW, OH WWW.MAGNA.EDU	MIDDLE 50% BETWEEN 21–26	MAJORITY IN TOP 50%	2.71	4-YR DEGREE	$8,500	$14,000	90%	35%

Student Information	Composite Score	Class Rank	Calculated GPA	Selected Major
	21	TOP 25%	3.29	ACCOUNTING

A dash (—) indicates information was not provided or could not be calculated. | *Institution provided cost information that may reflect more than tuition and fees.

College and Career Planning

Many people consider several possibilities before making definite career plans. Before you took the ACT, you had the opportunity to respond to questions about your educational and career plans. Use this information to consider possibilities that you may like to explore.

Interest Inventory Results

YOUR RESULTS INDICATE A PREFERENCE FOR WORKING WITH PEOPLE AND DATA

SEE MAP REGIONS 2, 3, 4
THE SHADED REGIONS SHOW CAREER AREAS HAVING WORK TASKS YOU PREFER.

RELATED CAREER AREAS:
COMMUNICATIONS & RECORDS
EMPLOYMENT-RELATED SERVICES
FINANCIAL TRANSACTIONS
MANAGEMENT
MARKETING & SALES
REGULATION & PROTECTIONS

College Major Selected

ACCOUNTING

THIS MAJOR PRIMARILY INVOLVES WORKING WITH DATA AND THINGS.

RELATED MAJORS:
BANKING & FINANCIAL SUPPORT SERVICES
BUSINESS ADMINISTRATION & MGMT, GEN
FINANCE, GENERAL
FINANCIAL PLANNING & SERVICES
INSURANCE & RISK MANAGEMENT
INVESTMENTS & SECURITIES
PURCHASING/PROCUREMENT/CONTRACTS MGMT
SMALL BUSINESS MANAGEMENT/OPERATIONS

Occupational Field Selected

INSURANCE & RISK MANAGEMENT

SEE MAP REGION 2
THE OCCUPATIONAL FIELD YOU CHOSE IS IN CAREER AREA C
MANAGEMENT

RELATED OCCUPATIONS:
ASSOCIATION EXECUTIVE
FINANCIAL MANAGER
FOREIGN SERVICE OFFICER
GENERAL MANAGER/TOP EXECUTIVE
HOTEL/MOTEL MANAGER
MANAGEMENT CONSULTANT
PROPERTY/REAL ESTATE MANAGER

The World-of-Work Map
(Your Interest Inventory results are shaded.*)

Four Basic Work Tasks: All college majors and occupations differ in how much they involve working with four basic work tasks: working with **People** (care, services), **Things** (machines, materials), **Data** (facts, records), and **Ideas** (theories, insights). These four basic work tasks are the compass points on the World-of-Work Map.

Regions and Career Areas: The map is divided into 12 regions, each with a different mix of work tasks. The map shows the locations of 26 occupational fields, called Career Areas (A–Z). Each Career Area contains many occupations that share similar work tasks.

*If no regions are shaded, you did not answer enough interest items to permit scoring.

For more information about your college and career planning, visit **www.actstudent.org** or check the booklet provided with this report.

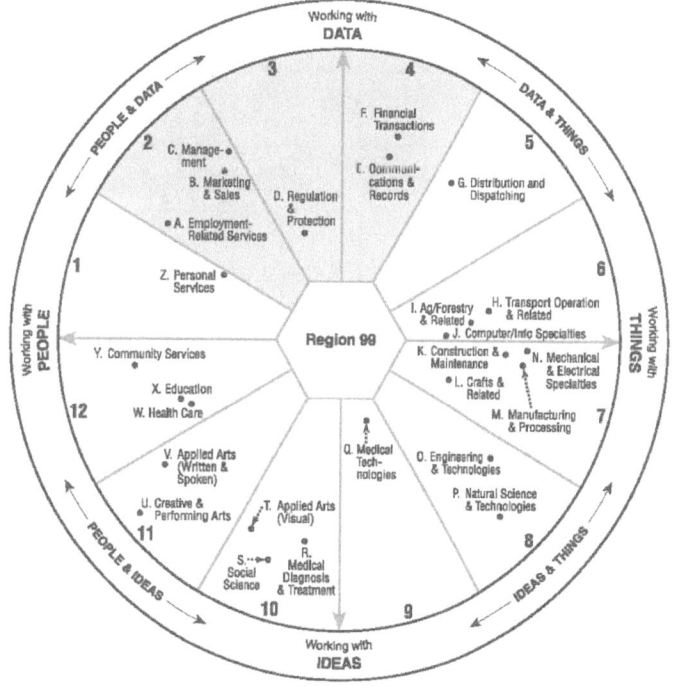

rSAT REPORT

SAMPLE STUDENT
SAT March 2016

Test Date: 2015 Nov 14

Total Score 950

Section Scores:	READING & WRITING	530	MATH	420
Test Scores:	READING	23	WRITING	30

Cross-Test Scores:	Analysis in History/Social Studies (H)	24	Science (S)	24

Subscores:

Reading&Writing		Writing or Math		Math	
Command of Evidence (CE)	7	Expression of Ideas (EI)	10	Heart of Algebra (HA)	7
Words in Context (WC)	9	Stnd Engl. Conventions (SC)	10	Problem Solving & Data Analysis (PS)	5
				Passport to Advanced Math (AM)	7

Essay (optional):   Essay-Reading **6**   Essay-Analysis **4**   Essay-Writing **6**

Reading

#	Std	Corr			#	Std	Corr		
1	B			.	31	C	D	5	
2	A	B		.	32	D	B	H	
3	B	C	WC		33	C	A	H	
4	C	A		.	34	B	C	H,	
5	D	C	CE					WC	
6	B	D		.	35	B		H	
7	A	D		H	36	A	D	H	
8	B		WC		37	B	C	H,CE	
9	C			.	38	C		H	
10	B		CE		39	A	B	H,CE	
11	D	A	H		40	B	C	H,	
12	A	B	H,					WC	
			WC		41	B		H	
13	B	D	H		42	A	B	5	
14	D	A	H,CE		43	A		5	
15	D	A	H		44	B	A	5	
16	C		H		45	C	D	5,	
17	D	C	H,CE					WC	
18	D		H,		46	D	C	5	
			WC		47	A	B	5	
19	A		H,CE		48	A		5,	
20	B		H					WC	
21	B	A	H		49	C	D	5	
22	B		5,		50	B		5	
			WC		51	D		5	
23	C	D	5,CE		52	D	A	5	
24	D		5						
25	C		5						
26	A	B	5						
27	D		5,						
			WC						
28	C		5,CE						
29	A		5,CE						
30	A		5						

Writing

#	Std	Corr		#	Std	Corr	
1	D		H,	31	C		EI
			WC,EI	32	A		SC
2	D	B	H,CE,EI	33	A		WC,EI
3	A		SC	34	A		EI
4	C		SC	35	B	A	WC,EI
5	C		H,EI	36	B		SC
6	A	D	H,CE,EI	37	D		CE,EI
7	B		SC	38	C		EI
8	C		SC	39	A		WC,EI
9	A		H,EI	40	B		SC
10	A		H,	41	B		SC
			WC,EI	42	A	C	CE,EI
11	B		SC	43	C	D	SC
12	A	B	S,CE,EI	44	B	D	SC
13	A		S,				
			WC,EI				
14	A	B	S,EI				
15	C		SC				
16	C		SC				
17	D	C	SC				
18	A		SC				
19	B	D	SC				
20	B	D	S,CE,EI				
21	B		S,				
			WC,EI				
22	C	D	S,EI				
23	D		WC,EI				
24	D		SC				
25	B		SC				
26	C	A	SC				
27	B		EI				
28	C		CE,EI				
29	B		CE,EI				
30	D		SC				

Math

#	Std	Corr	
1	D		HA
2	B	A	.
3	C		HA
4	B		HA
5	C		AM
6	A		HA
7	A	B	S,AM
8	B	C	AM
9	B		HA
10	D	A	AM
11	A	D	H,HA
12	B	D	HA
13	A	B	AM
14	C	A	AM
15	A	D	AM
16			AM
		2	
17			.
		1600	
18			HA
		7	
19			.
		0.8	
20			AM
		100	

Math

#	Std	Corr			#	Std	Corr	
1	C	B	PS		31	4.8		HA
2	C		PS		32	750		HA
3	A	D	.				107	
4	D	C	HA		33	625		H,PS
5	D		PS				0.63	
6	D		S,PS		34	1440		PS
7	C		H,PS				96	
8	D		HA		35	7		.
9	B	A	S,AM				6	
10	D	B	HA		36			AM
11	C	A	HA				3	
12	C		PS		37			H,AM
13	B	C	PS				1.02	
14	C		S,PS		38			H,AM
15	D	A	HA				6.11	
16	A	C	HA					
17	C	B	PS					
18	A		HA					
19	D	B	HA					
20	C	D	PS					
21	B	C	S,PS					
22	D	B	H,PS					
23	B		H,PS					
24	B	A	.					
25	A	D	S,AM					
26	C	B	S,PS					
27	A	C	S,PS					
28	A	C	HA					
29	D		AM					
30	C	D	AM					

BRAINSTORM TUTORING'S
HOW TO BEAT THE SAT/ACT MATH

{November 2015, updated 2019}

Beat the SAT/ACT Math Sections!
~The BrainStorm Method~

INTRODUCTION:

Math is a gas. Welcome to the good times that are SAT & ACT Math prep! There's much to learn. To be more specific: for the ACT, we need to re-learn some topics we haven't done in a while and add a few new, more advanced topics to our arsenal; for the SAT, we also need to dust off some old topics, but instead of also learning newer topics, what we need to learn is new approaches to math and, generally speaking, new ways of thinking in general. Either way, let's get this party started.

Get This Out of Your Vocabulary, Student:

   * "My math SAT score will definitely reflect my math grades in school."

   * "Since there's only one right answer to a math problem, there must be only one method to get to that answer."

Get This Out of Your Vocabulary, Tutor:

   * "Hey, you got the answer without explaining yourself or showing work—good job!"

   * "I'm good at math, so teaching it is easy."

   * "My sessions are so dynamite, no homework is necessary."

YOUR CHALLENGE, DEAR STUDENT:

Have you ever been to a foreign country and tried to speak the natives' language? Stumbled down tortuous alleyways, city map in hand, struggling to communicate with the local shopkeeper in a vain attempt to find your hotel? You should remember this clearly: though the architecture was magnificent, the views were stunning, and the city seemed (from an outsider's view) to be zipping along melodiously, you still felt like an outsider because you couldn't speak the language. If you understand the basics of a language—its grammar, its pronunciation, its format and presentation, even its history—you understand a culture... and as soon as you can do that, boom: you're an insider. You get it.

Math is a language in its own right—in fact, math, along with its sister language (music), is the world's universal language. Finding the slope of a line is the same process whether you're in Zurich or Macao, as is playing a G chord on a piano. In fact, most musicians are intensely gifted mathematicians and vice versa because each discipline is generated and catalyzed by the same part of the brain. In fact—wouldn't you know it—this part of the brain is also the place where human language resides and is processed! Those students of yours who are gifted and accomplished in one of these three areas—math, music or languages—are usually gifted in all three, and have only been developmentally hamstrung by incompetent teachers or lack of effort.

Your challenge is often to overturn your preconceived notion about yourself in relation to math. Being great at math is possible, it can be learned—and "you're just the person to learn it, my friend!" All we need is some structure and some creativity.

So here's the idea: teach/learn math like you would a foreign language to/with your cousin in Long Island. In no particular order, you should certainly focus on vocabulary, sentence structure, what's "to be avoided", etymology (a word's history), and applications to everyday life. You might teach/learn it by writing down vocabulary words on note cards, by drilling through examples and writing practice sentences, by employing the language regularly in both verbal and written media, by finding a pen pal (study partner), and, most important, by teaching it to someone else. Learning a language requires the world's greatest teacher–personal desire–and a certain level of immersion. Your job, Mr. or Ms. Student or Tutor, is to incite personal desire to succeed and to encourage yourself or your student by demonstrating that math is possible for anyone. You're a translator in this new city! You're a living map.

Pedagogy:

Let's review the idea of teaching math as a language. Here are some examples:

- **Vocabulary:** integer, improper fraction, negative number, product, reciprocal, mean
- **Sentence Structure:** $3x - 5 = 16... 6 + Z < -18$
- **"To be avoided":** zeros in the denominator, negative exponents, radicals in the denominator
- **Etymology:** Zero is a concept invented by the Arabs in 374 AD (what was life like without zero?!), trigonometry is Greek for "study of triangles", Pi was discovered in...
- **Practical applications:** How much tip do you leave the waiter? If you have a coupon to get bottled water at 20% off the original price, how much can you expect to pay now? How is batting average determined? If you drive at a certain rate, how long will it take before you're sipping piña coladas on the beach in Boca?

Once you've laid this conceptual/motivational groundwork, it's time to utilize BrainStorm's central learning technique: **Learn by Teaching Your Mentor**. How shall you teach/learn the math? With a mix of humor, drilling, vocabulary work, targeted quizzes, timed simulations, physical demonstrations, and reciprocal instruction, i.e., by teaching/learning the topics du jour & then exploring an assignment like this: "OK Kerry, you've done great today. Now, by the next time I see you, I want you to have reviewed these topics and prepared to teach them back to me. You should use props, drawings, example problems, etc. And I also want you to prepare me a quiz on these, and have the answers to your quiz at hand." Your tutor (or study partner) then can make you a quiz on the same concepts, and the next session is initiated by taking each other's quiz and tidying up the topics' loose ends before moving on.

Resources & Timeline:

This manual, along with its sister manual of practice quizzes later in this book, may act as your principal resource for test preparation. It should be supplemented with personalized quizzes as mentioned above, by any one of a number of SAT/ACT Math workbooks you may find at the local bookstore (we do not have a particular favorite to recommend), and by any one of the following websites:

www.majortests.com/sat/problem-solving.php

www.coolmath.com

www.onlinemathlearning.com

www.khanacademy.org

www.highschooltestprep.com

NOTE TO TUTORS: As for timeline, the first three steps, as outlined below (strategy, memorization, logic/manipulation), should take, at most, a couple of sessions. Depending on the student's time until test day, score goals, and current math competence, you will then proceed through the various topics outlined below. You should be able to cover ALL of the easy-to-medium-difficulty topics with your C.P. student in no more than 8 hours. In the same amount of time or less, you may be able to finish ALL the math topics (including those of high difficulty) with your more advanced students. Of course, the assiduousness and alacrity with which the student tackles his homework will determine as much as anything else how quickly you can move through the topics during sessions, so be sure to focus your energies on assigning clear, measurable, and ample homework—and making a bit of a stink when it has not been executed properly. The tolerance policy on slacking is ZERO TOLERANCE—the first time it happens, Mom & Dad know about it and everyone cracks down the following week. Do not underestimate this part of your job!!! [NOTE: I, Scott, typically assign my students 7-8 hours of homework per week and am very particular about HOW it should be executed.]

~OK, STUDENTS! LET'S GET ROLLING ON YOUR ACTION STEPS!~

THERE ARE SIX THINGS TO EXPLORE BEFORE YOU JUMP INTO YOUR CONTENT WORK...

1. STRATEGY
2. WHICH TEST?
3. MEMORIZATION
4. EQUATION LOGIC & MANIPULATION
5. THE "A THROUGH H" TECHNIQUE
6. THE ANALYSIS RUBRIC

1. STRATEGY.

There is much to say on this topic. I will try to keep it brief. For organization's sake, I have broken this section into two sub-sections: Intentionally Running Out of Time and The SAT/ACT Holy Trinity.

Intentionally Running Out of Time.

FIRST: introduce yourself to the basics of general ACT/SAT strategy: blanks, time management, & psychological management. Take either a full-length or at least a partial-length simulated exam during the week that you can analyze later. When doing the analysis, ask yourself: Am I focusing more on quality than on pacing? Am I allowing the clock to psych me out, and am I responding by doing rushed, poor-quality work? How many of my missed questions are due simply to sloppiness? To what extent did I allow external factors, such as noises, to distract me? Dig into the non-academic reasons you're struggling on the exam. Taking the test well is a skill that has nothing to do with math per se, and it requires sharpening.

SECOND: acknowledge these facts: the SAT Math score is composed of two sections for a total of 58 questions, which include 13 grid-in questions. Generally speaking,

students have about 90 seconds per question on the test—but truly have more like 100 seconds per question done with quality when we factor in the time some will save by intentionally not doing the (harder) questions at the end. The ACT Math score is derived from just one section of 60 questions in 60 minutes, giving us, strictly speaking, 1 minute per question, or something closer to 80 seconds per question done with quality (more on this in the next paragraph). The math sections all get into more advanced material as they go, so I constantly remind students: "Don't rush through the first two-thirds of the questions, doing a shoddy job because you're rushing, in order to have time on the harder questions at the end, which you probably will not get right anyway!" Don't be allergic to running out of time. It happens, it's ok, and you can get your goal score anyway! (unless you're going for a score above 700 SAT or 32 ACT, in which case: no running out of time for you!)

THIRD: we know that we should not be leaving blanks on the either test. This does NOT mean that every question is worth our time!!! Just like the SAT, the ACT requires the skill of "picking your battles"—most students can only do a great job on half to two-thirds of the math questions within the allotted time. Why are they all sprinting past these easier questions to labor away on the harder ones? This is a bad use of time, and a waste of quality work. On both tests, work backwards from a goal score to figure out how many answers you need to get right, and from there develop a pacing strategy that includes jumping questions (even if this means "leaving a guess answer behind" instead of leaving a blank). This is counter-intuitive, but spend MORE time on the easy-to-medium-difficulty questions to make sure you get them right—allow no unforced errors due to hasty reading or arithmetic! If you end up with enough time to tackle tougher questions at the end, fine. If not, no worries: you should have already hit your goal, so these extra questions are just gravy.

FOURTH: in terms of the effort to balance quality work with good pacing, many of my students have found value in splitting their test into sub-tests. Specifically, they change their mindset away from viewing the second SAT Math test as 38 questions in 55 minutes, away from viewing the ACT Math test as 60 questions in 60 minutes... and instead they chunk the test down into parts: the second SAT Math test is really a 12-question multiple choice test, followed by a 12-question multiple choice test, followed by a 6-question multiple choice test, followed by an 8-question grid-in test. Instead of one big test, they have 4 short tests. For each sub-test, the student assigns herself an amount of time for it: "Test 1, I have 15 minutes; Test 2, I have 15 minutes; Test 3, I have 10 minutes; Test 4, I have 15 minutes." The ACT version of this strategy is splitting the 60 questions thus: the first 30, the next 10, the subsequent 10, and the final 10. The timing is typically "Test 1, I have 24 minutes; Test 2, I have 12 minutes; Test 3, I have 12 minutes; Test 4, I have 12 minutes." Play with the concept of splitting long sections into sub-sections and see if it doesn't help you navigate the tenuous quality-pacing balance a little bit better.

 ## The SAT/ACT Holy Trinity.

Apart from sound pacing strategy, the holy trinity of strategy on the SAT includes the following:

1. Guess & Check and "Back Doors";
2. Annotation; and
3. Weaning off the calculator.

1. Many times we need to beat the SAT/ACT without using straightforward math formulas. This is where lateral brain and creativity need to kick in. For example, when both the questions and the potential answers are in the form of variables, why not plug in a number of your choosing and see where it takes you? If nothing else, it will often help the question to come alive and become conceptually cleaner—because as long as things remain in variable form, many students feel unduly mystified. "This question is impossible!" they grouse. "No, it's not!" you affirm. "It's just full of letters, which intimidates you!" Further, choosing your number can help eliminate a few answers that are clearly off-kilter. And never forget that you are taking a multiple-choice test; this means that the answer is right there in front of

you! You don't always have to approach the question "properly", i.e., from step 1 to step 5. You are often sage to use the "back door"—start from the answer option and work backward into the front of the problem. Ultimately, this first best practice is simply an encouragement to be resourceful—use every tool at your disposal to get yourself to an intelligent response.

2. Annotation is another killer strategy. I often say that the SAT Math test is, more than anything else, a reading test—did you note every detail in the question? Do you understand the vocabulary**? Are you paying attention to what the question is really asking in a big-picture sense?** Reading the question proactively—circling the question itself (Did they ask for the value of x? or of x^2? or of y?), focusing on comprehension as much as computation—helps to retain focus on the core requests that the question is making of you. I often have students go the end of the question & circle the goal outcome BEFORE doing any of the math itself. Avoiding small reading mistakes often is the difference between hitting one's overall goal scores and almost hitting that score.

3. Finally, the contentious issue of the venerated calculator. What students typically rely on to a fault in school mathematics—their 21st-century abacus—hampers them immeasurably on the SAT, which is a test designed to be performed without such assistance. The core competency the students need to evidence here is **number sense**: what happens when we divide by a fraction? How is it possible to figure out the 213th term in a sequence without counting? What is 14 times 23, roughly? We are sometimes to use number sense to eliminate certain options; for example, if the problem states that a variable in the denominator of a fraction increases, use what you know about number theory (a higher denominator means a lower value overall) to eliminate any answers that don't reflect this law. In other words, there are many cases in which good, levelheaded test-taking skills are the required competency—and calculator-driven number crunching is not. TI-84s do NOT take

the place of good, old-fashioned number sense, people! So put that thing away, student, and start re-familiarizing yourself with asking your brain to do math. Unless double-checking an answer, avoid the "siren call" of that calculator!

2. SAT vs. ACT: WHICH TO TAKE?

The CONTENT of the two tests overlaps by about 75%. That's a lot!! The STYLE with which that content is presented overlaps between the tests by about 50%. So by learning a number of the topics outlined below, you are truly preparing for both tests—but mostly on the level of content and not as much on the level of style/presentation. For example, an SAT problem might request the same thing an ACT problem requests—finding a point of intersection between two lines, for example—but the SAT might require the student to do so without a calculator AND to grid in the answer instead of choosing it from multiple-choice options.

In terms of the content that does NOT overlap between the tests, it mostly falls on the ACT side. That is, just about all the algebra, data, & geometry topics on the SAT are also on the ACT (albeit in a different form, as mentioned above), but the ACT includes far more non-planar geometry, as well as the following topics: logarithms, matrices, radian-degree conversion, and trig identities&graphs. Further, the ACT, while requiring knowledge of far more formulas than does the SAT, does not provide ANY of those formulas, so it behooves the student to make a big stack of note-cards as a way to practice quick recall of the dozens of formulas needed for the test. The ACT math section is also far faster than is the SAT math—another point in favor of SAT. But the SAT now has **an entire math section that is calculator free,** as well as 13 pesky grid-in questions. Plus, in truth: its questions are more convoluted, harder to interpret, thicker with variables and word problems. Hmm...

Here's an easy way to think of it: if you're pretty good at math in school; if you understand math when it's delivered in a straightforward manner; if you're good at remembering formulas; if you're tied to your calculator... the ACT might be good for you. If you're good at solving puzzles; if you love word problems and creative math; if you are comfortable doing math without a calculator... the SAT might be good for you.

So be your own coach: for which test should you prep? The answer in 90% of the cases is "Well, by prepping for one we're really prepping for the other, since a lot of the material overlaps. Let's take each test once (just a simulation, or the real deal), see how we've done, and then make a game plan from there as to which (if not both) we take again." Further, because students need not report their scores to their prospective colleges (this freedom to withhold score results from certain dates is called "Score Choice"), there is very little to lose by taking each test at least once, even if the majority of your prep is focused toward one or the other. In those cases in which it is glaringly obvious that you will be better suited for one or the other, then certainly: pick your horse and ride it into the sunset, leaving the other back at the stable. Otherwise, I generally have students take a simulation of each test and then, in some cases, take a real test of each as well. {NOTE: All colleges in the United States now accept either the SAT or ACT without preference. It's true! So do yourself a favor and really consider each one. No assumptions!!}

3. MEMORIZATION.

After exploring which test to take (if not both) and the relevant test goals and strategies, teach yourself this list of vocabulary terms: integer, proper fraction, improper fraction, whole number, rational/irrational number, real v. imaginary numbers, mean/average, range, mode, median, probability, direct/indirect proportionality (a.k.a. 'directly related' or 'inversely related'), sum, product, difference, quotient, parallel, perpendicular, circumference, perimeter, area, volume, reciprocal, prime number, factor, multiple, result, of, is, more than, less than, is subtracted from, ratio,

variable, constant, absolute value, consecutive, odd, even, distinct, "must be true" v. "could be true", ordered pair, complement, supplement, intersection, union.

Also to memorize: the prime numbers between 0-100, the most common Pythagorean triples (3-4-5, 5-12-13, 7-24-25, 8-15-17, 9-40-41, and multiples thereof), times tables 1-12, perfect squares up to 25^2, perfect cubes up to 10^3, and the formulas listed later in this SuperGuide.

The purpose of this exercise is both obvious and subtle. What is terrifically clear is that without the memorized formula, the student is all at sea when it comes to computing the area of a figure or the number of its diagonals. What is less obvious is that the valuable tool here is not JUST recall; it is also **speed of recall**. Taking 20 seconds to remember a certain formula is a waste of about 19 seconds, which of course adds up if it happens 8 times on a certain section. You need to have these formulas down cold—and I mean shiver-down-your-spine, "damn, that dude is a robot!" cold. This is a matter of effort, not talent, so let's not make excuses for being fuzzy on formulas or definitions. Instead, let's make it automatic and fast, so that our creative energies can be properly focused where they need to be: the question's concept and execution.

4. EQUATION LOGIC & MANIPULATION.

The central logic of equations is that they can be manipulated. You can add 7 to both sides, you can divide both sides by -13, you can multiply both sides by the variable M, etc. This theory is helpful to keep in mind when you're considering a geometric problem, for instance. If the problem asks for the area of a circle, **you should first write out the formula** (most lazy, untrained students do not do this simple, sky-opening discipline); second, you should begin an internal dialogue: "Hmm. Since the formula contains two variables, it must be my job to figure out 'r' so that by plugging it in I can get my goal answer of 'A,' area. Thus, now I know that the question is really secretly asking me to find 'r'. Let's go get him!" This "working backwards" logic is a seminal virtue to solving math word problems, and should be learned earlier rather than later.

So too should equation manipulation. Almost every equation at this level can be rewritten in several different ways, & you should be familiar with this skill. For example, if it's true that Average = Sum of Terms/Number of Terms {or A = S/N}, it follows that S = AN and N =S/A. It's the same exact equation written three different ways! But do you not have the ability to manipulate the memorized formula? It should not be! Make sure you are aware that the test will assess your ability to use the concept of average—or rate & distance, or exponent rules, etc—in a variety of ways, so you must feel perfectly comfortable with this sort of manipulation.

Think of a few examples of this concept that "although it looks different, we are dealing with the same thing here." Here's one: there is no difference between the improper fraction 5/2 and the mixed fraction 2 ½ other than how they look. Similarly, there is no difference among the fraction ¼, the decimal 0.25, and the percentage 25%. Or for verbal types, there is absolutely no difference, grammatically speaking, between the following two sentences—except for how they look:

Under the table are my gloves.
My gloves are under the table.

So what if you did the work on the problem and you got an answer of 25.13, but you don't see it listed? Can't you see that Option B says 8π, and that this is identical to your decimal answer? Time to get used to things showing various manifestations but meaning the same thing.

5. THE "A THROUGH H" TECHNIQUE.

Some students whiz right through math sections, no problem, and ace the test. If you're one of them, congratulations! Jump right ahead to the content section later in this book.

If, however, you're like most people, you sometimes miss math questions. And sometimes you miss those questions not because you don't know the material, but because you made an UNFORCED ERROR. Ahhhh!! Yikes. How frustrating to miss questions that you knew how to do.

I'll say that an average student on the ACT Math section misses 5-6 questions due to sloppy mistakes ranging from mis-reading the question to doing poor calculations. That's a big deal! That's the difference between a 19 and a 22, or the difference between a 29 and a 33! HUGE!

Typically, students make those mistakes because of bad habits. To undo those habits, we need to install a new set of steps to attack each question. WARNING: This process is intentionally annoying. It is laborious; it feels unnecessary, and it by definition slows people down. But as mentioned above, you have the choice between two things: keeping your old habits, thereby sacrificing a handful of very doable questions you should really get; and creating a new habit that, yes, is cumbersome, but YES, is worth it!! NOTE: play with all 8 steps. You may end up using all of them as described below, or you may choose just a few of the steps to keep yourself on track. You can certainly customize this process to your needs.

THE A-H TECHNIQUE gives you an 8-step journey to help you navigate the questions of the SAT/ACT Math. Don't forget that this process is not only to be used on "hard" or "advanced" questions. Its whole purpose is to help you avoid the pitfalls of sloppy reading and calculating that plague us all– especially on the "easier" questions.

Without further ado, here she is!

A – ANSWERS FIRST

B – BOX THE QUESTION

C – CLUES UNDERLINED

D – DUPLICATE

E – ENSURE

F – FINALLY, MATH

G – GUARANTEE

H – HOORAY!

Explanation:

A – ANSWERS FIRST. Before reading the question itself, do a quick 2-second skim of the answers. Are they fractions? Are they variables? Let them give you a clue as to the ultimate destination of the problem so your subconscious can start working on it.

B – BOX THE QUESTION. Next, work your way up to the very last sentence of the problem and draw a box around the question itself. Find the crucial term: median. maximum. Y. Don't miss it: box it.

C – CLUES UNDERLINED. Now read the question, underlining (annotating) as you go. Every single clue word or equation or number gets an underline. If the phrase is "five consecutive odd integers", you underline every single one of those words.

D – DUPLICATE. Don't start working on an equation until you've written it out. Again, this sounds tedious, but do it. If you are given a big, strange equation, re-write it exactly in the margin. If they imply the need for an equation, such as the sum of the interior angles of a hexagon, go to the trouble of writing the equation $S = 180 (N - 2)$ in the margin. Note that a formula/equation has an equals sign with terms on both sides. Don't just write "$180 (N - 2)$." Write the whole thing. And "duplicate" any other crucial clue you're given by writing it in the margin as well. Re-writing it forces your brain to truly acknowledge something and pass it by.

E – ENSURE. Confirm that everything you've written in the margin is perfectly duplicated and that you've missed no clues. You know how to proceed and you've missed nothing so far.

F – FINALLY, MATH. Now, you finally get to jump into the calculating you wanted to leap into from the get-go. Only now you've set it up properly and haven't missed any details. Notice also that you need to do the math carefully by writing out only one step at a time~ no jumping steps!

G – GUARANTEE. If time permits, look at your answer. Before moving on from the problem, guarantee that the answer is correct by plugging it back into the original equation, or by getting the same answer a different way.

H – HOORAY! You did a great job! Give yourself a quick "You rock!" Before moving on, put a check mark next to the number on the page, indicating to yourself that it's good to go.

Working through each problem this way may slow you down at first, but it will actually speed you up in the long run. Do it in faith until it becomes smooth.

One more bonus tip: don't fill your answer into the scantron sheet immediately after finishing the problem. Do ALL the questions on the two pages in front of you (left and right) and THEN fill in all of the answers in one fell swoop. This will save you time.

6. THE ANALYSIS RUBRIC.

Let's say you've taken a practice ACT Math section and graded it. You got 37 correct and 23 incorrect. Now what?

The aim to learn something that will serve you next time you take the test. To do so, I suggest using the following analysis rubric:

> **1** – Put a '1' next to any wrong answer if you missed the question due to an UNFORCED ERROR.
>
> **2** – Put a '2' next to any wrong answer if you missed the question due to NOT KNOWING A FORMULA OR VOCABULARY TERM.
>
> **3** – Put a '3' next to any wrong answer if, after reading the book's explanation or hearing a tutor explain it, you totally see it and realize you could have gotten it, but just didn't connect the dots during the test. We call these questions NOW I SEE IT.
>
> **4** – Put a '4' next to any wrong answer if it's truly baffling to you. If it's way advanced and you have no clue what's going on when you read the explanation, you rate the question a '4'. These I call NO IDEA questions.

Now, tally up each category. How many 1's were there? How many 3's? Notice the primary cause of your wrong answers and address it. If you missed 7 questions because of unforced errors, shoot! Time to really start working the A-H

Technique! If you missed 4 questions because you didn't know such terms as "irrational number" or such formulas as that for percent change, or if you forgot your prime numbers, etc., then get working on your memorization! If you missed a good deal of NOW I SEE IT problems, great— keep taking practice sections and you'll start seeing things faster. If you missed a good number of 4's, well: you need to work on your content knowledge.

      ALWAYS FOCUS FIRST ON ELIMINATING 1's AND 2's. The 3's and 4's will follow as you keep chipping away at your content knowledge and taking practice tests.

OK great~ we did it! We covered a lot of bases there.

Now, my friend, you are ready...

DRUMROLL, PLEASE...
IT'S TIME FOR THE

INTRODUCTION OF TOPICS!!

Whew! OK—so far so good. You've learned general strategy of the test, strategy specific for the math sections, all of the key vocabulary & formulas, and the skills of logic & manipulation. You've learned the "A through H" method and understand the analysis rubric. Maybe you've even figured out which test you're taking. You're ready to introduce the topics! The topics are listed below & elaborated on thereafter.

FIRST LEVEL OF TOPICS:
- Solving basic equations and inequalities;
- Finding mean, median & mode from a set of data (whether from a set of numbers or a chart/graph);
- Finding the distance & the midpoint between two points in a Cartesian plane;
- Directly/inversely related word problems;
- Angle relationships with parallel lines;
- English to Algebra translation problems;
- LCD and GCF;
- FOIL and PEMDAS;
- Arithmetic Algebra;
- Basic percent problems

NEXT LEVEL OF TOPICS:
- Solving systems of equations;
- Powers and roots (i.e., exponent and radical rules);
- Finding the slope & equation of a line;
- Slopes of parallel & perpendicular lines;

- Functions (equations & basic graphs);
- Absolute value in equations;
- Finding angle values, side values, number of diagonals, etc., in polygons;
- Radius/Diameter/Circumference/Area of circle;
- Basic Triangle rules (Triangle Inequality Theorem, Pythagorean Theorem, Area);
- Factoring (Difference of Two Squares, Trinomials);
- Properties and formulas of various polygons;
- Radian-Degree conversion {ACT only};
- Internal/external angles and diagonals of polygons

UPPER LEVEL I: PROBLEMS THAT REQUIRE CREATIVITY/MEMORIZATION:
- Ratios {A:B or A:Total};
- Percent increase & decrease;
- Inscribed shapes & shaded area;
- Number theory [the vocab-heavy roman numeral questions];
- Created symbol functions;
- (Rate)(Time) = Distance;
- Special right triangles;
- Similar triangles;
- Word problems involving average;
- Word problems involving consecutive integers;
- Word problems requiring Venn diagrams;
- The sum of a large set of terms;
- Number line questions;
- Total Cost = Fixed Cost + Variable Cost;
- The nth term of an iterative sequence;
- Standard equation of a circle in the coordinate plane;
- Quadratic formula;

- Counting Principle, Combinations;
- SohCahToa and Law of Sine & Cosine;
- Word problems of the form "If Ken can mow h lawns in m hours, how many minutes will it take him to mow j lawns?" (Advanced Ratio)

UPPER LEVEL II: TOPICS THAT NOT ALL STUDENTS HAVE DONE YET IN SCHOOL:
- Sector area & arc length in circles;
- Graph translations & transformations;
- Composite and inverse functions;
- Absolute value in inequalities;
- Arithmetic and Geometric sequences;
- RT=D involving conversions or asking for average rate over two trips;
- Geometry questions involving radicals, fractions, and variables;
- Combining two ratios with a common term;
- Graphing Inequalities;
- Matrices {ACT only};
- Trig identities & graphs {ACT only};
- Logarithm-Exponent Equation conversion {ACT only};
- Imaginary numbers;
- Ellipses & hyperbola;
- Unit circle

Conclusion:

The SAT/ACT math often requires 12-15 hours of personal effort + 8 hours of tutoring (or about 25 hours of personal effort without tutoring) to achieve really solid growth. And like the grammar section, it is composed of topics that you can run through one at a time like a checklist.

Pick your battles. If you aren't afforded all the time you'd like to get through all the material, choose those topics you feel are most relevant and most quickly learned. **I have found that introducing sound strategy & time management based on specific score goals is the fastest way to catalyze increase**; thereafter, solidifying vocabulary, equation logic/manipulation, & basic equation solving is most effective. If time permits, you can also jam through the three levels of topic difficulty, as outlined above.

A note on homework: do it. Don't be shy. This is a language you are trying to learn, remember? And language requires constant (read: daily) reinforcement. Even if it's only 20-25 minutes a day of memorization, drilling, & problem solving, you will greatly benefit from consistent rigor. Soon you will find your fluency and confidence improving. And once you've gotten to this point, you've done it! You've prepared yourself to

STORM THE TEST!

55

NOTE: The best math teachers do not assume that their students know the basics of math. Instead, they go right through the most rudimentary of topics one by one, making sure there are no small gaps that lead to huge holes in more advanced problems. These topics—many organized and articulated by a couple of verbal studs who needed to get a handle on teaching math, their nemesis—are a good starting point for reviewing basic rules & formulas.

- **Solving basic equations & inequalities.**
  - Linear equations: follow two basic rules to isolate the variable of interest:
    - Perform the OPPOSITE operation to what is currently being presented in the equation {if the expression on the left of the equation says "5x – 9", we notice a subtraction taking place as well as a multiplication; to isolate the x we will therefore add (the opposite of subtraction) and then divide (the opposite of multiplication).
    - Perform the action to BOTH sides of the equation.
  - Ex: $5x - 12 = -2x + 9$

    $5x + 2x = 9 + 12$

    $7x = 21$

    x = 3 {p.s.—don't forget to check the answer by plugging it back into the original equation!}

- **Quadratic equations.** Put the equation into "$ax^2 + bx + c = 0$" form, then factor the left side & set each parenthetical factor = 0 to get two solutions.
    - Ex: $x^2 + 12 = 7x$

      $x^2 - 7x + 12 = 0$

      $(x - 3)(x - 4) = 0$

      $(x - 3) = 0$ OR $(x - 4) = 0$

      $x = 3$ OR $x = 4$
    - NOTE: Remember that if the "$ax^2 + bx + c$" trinomial is un-factorable, we need to use the quadratic formula (below). Remember that what is under the radical {"$b^2 - 4ac$"} is called the determinant, because it determines what KIND of answers we will get. That is, if, once plugging in for the variables, the term's value is positive, we can expect to get 2 real answers for x. If the term's value is zero, we can expect 1 real answer. If, however, the term's value is negative (for example, when b=3, a=5, and c=11), we can expect 2 imaginary answers. This means that if we were to graph the function, we would see a parabola that never intercepts the x-axis.

$$x = \frac{-b \pm \sqrt{b^2 - 4ac}}{2a}$$

- **Equations with fractions.** Find LCD in denominator, then multiply it by each number in the equation.
    - Ex: $^m/_3 - {}^m/_4 = 2$

      $(12)\,^m/_3 - (12)^m/_4 = (12)2$

      $4m - 3m = 24$

      $m = 24$

- **Cross multiplication.** Used when one fraction equals another fraction.

  o Ex: $\frac{x}{3} = \frac{x+4}{5}$

  $5x = 3 (x + 4)$

  $5x = 3x + 12$

  $2x = 12$

  $x = 6$

- **System of equations.** Solve for two variables by combining the equations so one of the variables cancels out—a procedure called "elimination."

  - Ex: $4x + 3y = 8$ AND $x + y = 3$ (multiply each side of this equation by -3)

    $-3x - 3y = -9$

    $x = -1$ (now plug in to one of the original equations)

    $4(-1) + 3y = 8$

    $-4 + 3y = 8$

    $3y = 12$

    $y = 4$

    The final answer is reported as the ordered pair (-1, 4). This ordered pair represents the point at which the two lines intersect.

  o NOTE: What happens if, after lining up the equations for elimination, you find that BOTH the x and y terms will cancel, leaving you with 0 on the left side of the equation after you add? One of two things: if the numbers on the right side of the equation also cancel, leaving you with 0 on the right, what we have is $0 = 0$, which is of course always true. Meaning: the two equations represent the same exact line, which means they intersect at infinitely many points. Second scenario is that, after adding the numbers on the right, you are left with a number other than 0 (for example, you're left with $0 = 7$, which is never true).

Meaning: the two equations are different lines that never intersect, i.e., they are parallel.

- **Inequalities.** Do whatever is necessary to both sides to isolate the variable.
  - REMEMBER: when you divide or multiply both sides by a negative number, reverse the inequality sign! Otherwise, all regular rules for isolating the variable apply (exceptions to be discussed later: absolute values and exponents in inequalities).
  - Ex: $-5x + 7 < -3$

    $-5x < -10$

    x > 2

- **Finding Mean, Median. & Mode from a data set.**
  - Average: Add numbers and divide sum by amount of numbers.
    - $avg = {}^{\text{sum of values}}/_{\text{number of values}}$
  - Algebra Average: sides of a triangle:
    - Ex: $2x + 1, x + 7, 3x - 11$

      $$\frac{(2x + 1) + (x + 7) + (3x - 11)}{3} = \frac{6x - 3}{3} = \textbf{2x - 1}$$
  - Finding an unknown number when average is given:
    - $S = AN$, as mentioned on page 6 of this packet.
      - Ex: Average = 78. 3 #'s in set are 71, 74, 83. What is the 4th number?

        $78 \times 4 = 312$ (since avg is 78, $^{312}/_4 = 78$)

        Σ of 3 #'s = 228

        $312 - 228 = \textbf{84}$

- Ex: Average of w, x, y, z = 31. If avg of w + y = 24, what is avg of x + z? (Use S = A*N)

    w + x + y + z = (31) 4 = 124 AND $\frac{w+y}{2}$ = 24 ⇒ w + y = 48.

    Substitute 48 for w + y: x + z + 48 = 124

    x + z = 76

    avg of x + z = **38**

o Weighted Average: average of 2 or more sets:
  - multiply avg of each set by # of values in that set
  - then add the products
  - then divide the sum of the products by the total # of values in the sets.
    - Ex: 18 kids have an average midterm grade of 85. 12 kids have an average midterm grade of 90. What is the average grade of the entire class? (30 kids)

      <u>Set 1 (Average x Amount) + Set 2 (Average x Amount)</u>
      Total # of Students

      $\frac{(85 \times 18) + (90 \times 12)}{30}$ ⇒ $\frac{1530 + 1080}{30}$ = **87**

o Median: arrange #s in increasing order. If a set contains an odd # of values, the median is the middle value. Ex: 8, 12, 15, **17**, 19, 20, 25
  If a set contains an even # of values, median = avg of 2 middle values.
  - Ex: 10, 20, **24, 30**, 40, 50 ⇒ (24 + 30)/2 = **27**

o Mode: # that appears the most. Ex: 7, 2, **6**, 3, 2, **6**, 7, 3, 9, **6** ⇒ Mode = **6** {NOTE: If there are two or more numbers that appear an equal amount of times, we do NOT take the average of these modes for one overall mode of the set (as we do with median). Instead, the set simply has more than one mode.}

- **Finding distance between 2 points; i.e, Distance Formula.**
    - $$D = \sqrt{(x_2 - x_1)^2 + (y_2 - y_1)^2}$$

        - Ex: (3, 6) and (5, -2) : $d = \sqrt{(3-5)^2 + (6 - -2)^2}$
            $$\sqrt{-2^2 + 8^2}$$
            $$\sqrt{4 + 64}$$
            $$\sqrt{68} \Rightarrow 2\sqrt{17}$$

    - NOTE: This formula is simply another way of writing the Pythagorean Theorem, as the distance between two points is simply the length of the hypotenuse of a right triangle.

- **Directly Proportional Word Problems.** Two values are directly proportional to each other when the ratios are constant. Conceptually, this means that the two values move in tandem, i.e., in the same direction.
    - Ex: If 2 pencils cost \$1.50, how many can you buy with \$9.00? (# of pencils is directly proportional to the cost).

        $\dfrac{2}{1.50} = \dfrac{x}{9.00} \quad \Rightarrow 18 = 1.50\,x \Rightarrow x = 12$

- **Inversely Proportional Word Problems.** When the product of two values is constant. Conceptually, this means that the two values move in opposition, i.e., when one goes up the other goes down.
    - Ex: It takes 4 men 6 hours to repair a road. How long will it take 8 men to do the job at the same rate? (# of men is inversely proportional to the time taken to do the job).

        $4 \times 6 = 8 \times t \Rightarrow 8t = 24 \Rightarrow t = 3$ **hrs** o   Ex: The number of Chihuahuas varies inversely with the number of piranhas. When there are 30 Chihuahuas there are 40 piranhas. How many Chihuahuas are there when there are 120 piranhas?

30 x 40 = C x 120 ⇒ **C = 10 Chihuahuas**

- **Angle Relationships with parallel lines.** Always the same distance apart and do not intersect.
    o   The acute angles are supplementary to the obtuse angles (i.e., they add to 180).
    o   Intersecting lines: adjacent angles are supplementary, vertical angles are equal. Remember: if we get the measure of one angle, we can get the measure of all the other angles around the same transversal. SPECIAL NOTE: Beware! When two different transversals are drawn through the same parallel lines, do not get mixed up. Cover one transversal and get all the angles of the first; then cover the first and get the angles of the second. They should not be intermingled or confused as somehow related to each other!!!

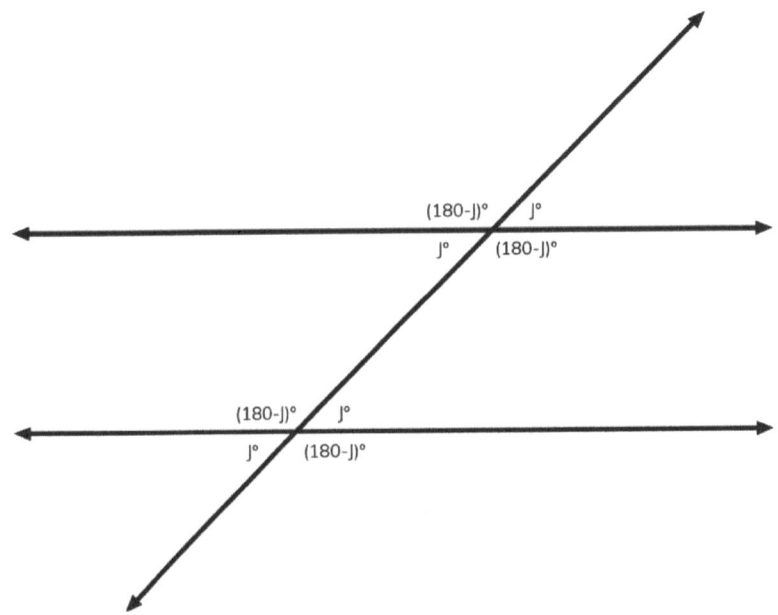

- **English to Algebra Translation.** Write out algebraic expressions as you read them.
    - o Ex: "Three more than twice a number is four less than one-half the number."

        Every word turns into an algebraic expression. It becomes

        $3 + 2x = \frac{1}{2}x - 4$ (notice that "4 less than" means "subtract 4")

        $6 + 4x = 1x - 8$ (we have multiplied through by the LCD)

        $3x = -14$

        x = -14/3
    - o Ex: "40 is what percent of 25." This becomes, through direct translation,

        $40 = (P/100) \times 25$ (notice that "percent" means "over 100")

        $8/5 = P/100$ (we have divided by 25 and reduced)

        $5P = 800$ (we have cross-multiplied)

        P = 160%

- **FOIL: First, Outer, Inner, Last (to multiply binomials)**
    - o Ex: $(x + 3)(x + 4) \Rightarrow x^2 + 3x + 4x + 12 \Rightarrow$ **$x^2 + 7x + 12$**
    - o Ex: $(5 - 2i)(6 + 5i) \Rightarrow 30 + 25i - 12i - 10i^2$

        We know that $i^2 = -1$ (see formulas for imaginary numbers), so we can substitute -1 for the i^2 term: $30 + 25i - 12i - 10(-1) \Rightarrow$ **$40 + 13i$**
    - o Special Pairs of Binomials:
        - ▪ $(a - b)(a + b) = a^2 - b^2$ (this is called "the difference of two squares")
        - ▪ $(a + b)^2 = (a + b)(a + b) = a^2 + 2ab + b^2$
        - ▪ $(a - b)^2 = (a - b)(a - b) = a^2 - 2ab + b^2$

- **PEMDAS: The acronym for order of operations.** {Parenthesis, Exponents, Multiplication, Division, Addition, Subtraction}
    - Ex: $3 - 9[4 - (-5)]^2 \Rightarrow 3 - 9[9]^2 \Rightarrow 3 - 9(81) \Rightarrow 3 - 729 \Rightarrow$ **-726**

- **Exponent Rules (no calculator allowed!).**

    $9^0 = 1$ (anything to the zero power is 1)

    $9^1 = 9$ (anything to the first is itself)

    $9^2 = 81$

    $9^{-2} = \dfrac{1}{9^2} = \dfrac{1}{81}$ (a negative power means to take the reciprocal of the base)

    $9^{1/2} = \sqrt{9} = 3$ (a fractional power means to take the root of the base)

    $9^{-1/2} = \dfrac{1}{\sqrt{9}} = \dfrac{1}{3}$

    - $(x^2)(x^3) = x^{2+3} = x^5$ (when multiplying equal bases, add exponents)
    - $(x^4)^3 = x^{4(3)} = x^{12}$ (when raising a power to another power, multiply)
    - $(x^{1/3})^{4/5} = x^{(1/3)(4/5)} = x^{4/15}$
    - $x^4 + x^3 = x^4 + x^3$ (when adding/subtracting, must have same base AND same power!)
    - $3x^5 - 7x^5 = -4x^5$
    - $\dfrac{x^4}{x^3} = x^{4-3} = x^1$ (when dividing equal bases, subtract exponents)

- **Fraction Rules (no calculator allowed!).**
    - Reducing Fractions: $\dfrac{28}{36} = \dfrac{4 \times 7}{4 \times 9} = \dfrac{7}{9}$
    - Adding & Subtracting (requires LCD):

        $\dfrac{2}{15} + \dfrac{3}{10} = \dfrac{4}{30} + \dfrac{9}{30} = \dfrac{4+9}{30} = \dfrac{13}{30}$
    - Multiplying (multiply across): $\dfrac{5}{7} \cdot \dfrac{3}{4} = \dfrac{5 \times 3}{7 \times 4} = \dfrac{15}{28}$
    - Dividing (multiply by the reciprocal of the second fraction):

        $\dfrac{1}{2} \div \dfrac{3}{5} = \dfrac{1}{2} \cdot \dfrac{5}{3} = \dfrac{5}{6}$

- o Mixed Numbers: multiply the whole number by the denominator and then add to the numerator: $7\frac{1}{3} = (7 \times 3) + 1 = \frac{22}{3}$
- o Improper Fractions: divide the denominator into the numerator to get a whole number with a remainder: $\frac{108}{5} \Rightarrow 108 \div 5 = 21\ r\ 3 \therefore \mathbf{21\frac{3}{5}}$
  - ▪ Ex: $\dfrac{6}{x\text{-}3} + \dfrac{x\text{-}2}{x+4} = \dfrac{6(x+4) + (x\text{-}2)(x\text{-}3)}{(x\text{-}3)(x+4)} = \dfrac{6x + 24 + x^2 - 5x + 6}{(x\text{-}3)(x+4)}$

$$= \dfrac{x^2 + x + 30}{(x\text{-}3)(x+4)}$$

- ▪ **Arithmetic Rules (no calculator allowed!).**
  - o $2 + 2 = 4$
  - o $-2 + 2 = 0$
  - o $-2 - 2 = -4$
  - o $(-2)(-2) = 4$
  - o $2^2 = 4$
  - o $(-2)^2 = 4$
  - o $(2)^{-2} = \frac{1}{4}$
  - o $2 - (-2) = 2 + 4 = 6$

- ▪ **Number Theory.**
  - o This topic is all about number sense. We can test our own numbers, but that can lead to error—how do you know, for example, that another number would have given you a different result? This is a problem mostly heavy favored on the SAT, and has to do with speaking through number properties.
  - o Ex: If $x < -1 < y < 0$, which of the following MUST be true?

    I. $y - x > 0$ [Well, if I take a negative fraction & subtract a more negative number, yes, the answer will be positive, no matter what

example numbers I choose, because subtracting a negative is like adding a positive. **YES.**]

II. yx > 0 [Well, if I multiply two negative numbers together, the result is always positive, regardless of the numbers I choose here. **YES.**]

III. y^x > 0 [Well, a negative exponent flips the base, which will make y an improper negative fraction. If x is an even number, it will make the expression positive, because any base to an even exponent goes positive. So this problem COULD be true. But if x is an odd number, it will keep y negative. So the answer here is **NO.**]

- **Square Root Properties.**

  - $\sqrt{12} = \sqrt{4x3} = \sqrt{4} \times \sqrt{3} = 2\sqrt{3}$
  - $2\sqrt{3} + 3\sqrt{3} = 5\sqrt{3}$ (with addition & subtraction, numbers in radicals have to match)
    - $a\sqrt{b} + c\sqrt{b} = (a+c)\sqrt{b}$
  - $\sqrt{4+9} \neq \sqrt{4} + \sqrt{9}$
    - $\sqrt{a+b} \neq \sqrt{a} + \sqrt{b}$
  - $\sqrt{3} \times \sqrt{5} = \sqrt{3x5} = \sqrt{15}$
    - $\sqrt{a} \times \sqrt{b} = \sqrt{axb} = \sqrt{ab}$
  - $\dfrac{\sqrt{6}}{\sqrt{3}} = \sqrt{\dfrac{6}{3}} = \sqrt{2}$
    $(\sqrt{5})2 = \sqrt{5} \times \sqrt{5} = 5$
  - eliminate $\sqrt{}$ from denominator (called "rationalization"):
    - $\dfrac{5}{3} \times \dfrac{\sqrt{3}}{\sqrt{3}} = \dfrac{5\sqrt{3}}{3}$

- **Finding the slope and equation of a line.**
  - slope = $\dfrac{\text{change in y}}{\text{change in x}}$ = $\dfrac{\text{rise}}{\text{run}}$ = $\dfrac{\text{vertical change}}{\text{horizontal change}}$

- A (2, 3), B (0, -1) : $\dfrac{y_a - y_b}{x_a - x_b} = \dfrac{3 - (-1)}{2 - 0} = \dfrac{4}{2} = 2$

  - lines with the same slope are parallel.

  - lines with negative reciprocal slopes are perpendicular.

- Equation of a line, slope intercept form: $y = mx + b$, where m represents the slope and b represents the y-intercept. To get the x-intercept, plug in 0 for y and solve for x.

- Equation of a horizontal line is always in the form $y = k$, as in $y = 5$. Slope of any horizontal line is 0.

- Equation of a vertical line is always in the form $x = k$, as in $x = 3$. Slope of any vertical line is undefined.

- Standard equation of a line: $Ax + By = C$. This is really just a reformatting of the slope-intercept format, in which the "mx" term has been brought to the left side and all fractions have been eliminated by multiplying through by an LCD.

- **Functions. {problems involving "f(x)", said "f of x"}**

  - NOTE: These problems, generally speaking, are simply plug-in problems. The equation includes an input (x) and an output (y, or f(x)). All we have to determine is which we've been given, plug into the appropriate place, and solve.

  - $f(x) = 2x + 3$. Find f(4). (We have been given x, so replace all x's with 4)

    $f(4) = 2(4) + 3 = \mathbf{11}$

  - $f(x) = 3x^2 - 48$. Find x if f(x) = 0.

    In this case, we have been given f(x), a.k.a. y. So we plug the 0 in for f(x) and get

    $0 = 3x^2 - 48$

    $0 = 3 (x^2 - 16) \Rightarrow 0 = 3 (x - 4)(x + 4) \Rightarrow \mathbf{x = 4, -4}$

- o Composite functions are the in-laying of one function into another. The key rule is WORK FROM THE INSIDE OUT.
  - Ex: If $f(x) = x + 1$ and $g(x) = 7 - 4x$, find $g(f(-3))$.

    First we'll plug the x of -3 in for x into the $f(x)$ equation, so

    $f(-3) = -3 + 1 \Rightarrow f(-3) = -2$

    Now we'll plug our -2 into the $g(x)$ equation, so

    $g(-2) = 7 - 4(-2) \Rightarrow g(-2) =$ **15**
- o The domain of $f(x)$ is all possible values of x; all the values that x is legally allowed to be based on mathematical common law.
  - Ex: In $f(x) = \frac{1}{x}$, x cannot equal 0, as it is illegal for 0 to be in the denominator; $f(x) = 1/(1-x^2)$ excludes 1 and -1 because these numbers would make the denominator zero
  - Ex: In $g(x) = \sqrt{x-3}$, $x \geq 3$, as anything below a value of 3 for x would give us a negative number under the radical, which would give us an imaginary number, which we do not like.

- **Absolute Value.** Definition: the distance of a number from 0 on the number line. Because it is distance, the result is always POSITIVE. Every positive number is the absolute value of two numbers: itself and its negative.
  - o $|7| = 7$; $|-7| = 7$
  - o If $|x| = 13$, then **x = 13, -13**
  - o **Absolute Value in Inequalities: the case of the ballerina and the bodybuilder.** If the variable is within the absolute value symbol, and this symbol, after being isolated in the inequality by moving everything over to the other side, is LESS than the number on the other side, you have a ballerina. If the absolute value is GREATER than the number on the right, you have a bodybuilder. Meaning: in the case of the former, the answer should a finite set of numbers on either side of zero (an answer along the lines of "-4<x<6"); in the case of the latter, the answer

should be two infinite sets of numbers reaching away from 0 (an answer along the lines of "-11<x or x>3"). When you graph the first, the two endpoints reach toward each & meet in the middle, like a ballerina's hands; when you graph the second along the number line, it looks like two arms flexing and hands reaching outwards ("which way to the gym?").

- Ex: $-4|x + 9| < -24 \Rightarrow |x + 9| > 6$ (notice that we divided by a negative so we had to flip the inequality sign)

  Now we split into two inequalities: $x + 9 > 6$ OR $x + 9 < -6$

  Solve each: $x>-3$ OR $x<-15$. Final answer: **-15<x OR x>-3**

- Ex: $3|x - 2| + 7 < 19 \Rightarrow 3|x - 2| < 12 \Rightarrow |x - 2| < 4$

  We now split it into two inequalities: $x - 2 < 4$ AND $x - 2 > -4$

  Solve each: $x < 6$ AND $x > -2$. Final answer: **-2<x<6**

- **Basic Geometry.**
  - Circles.
    - RADIUS (r) = line from center of circle to point on the circle's circumference
    - DIAMETER (d) = line thru center of circle (longest line); $d = 2r$
    - CIRCUMFERENCE = "perimeter" of circle; $C = 2\pi r$ or $c = \pi d$
    - AREA = πr^2
    - Equation of a circle on a line, where r signifies radius and (h,k) signifies the center: $(x-h)^2 + (y-k)^2 = r^2$
      - Ex: In equation $(x+3)^2 + (y-6)^2 = 20$, the center is (-3, 6) and the radius is $\sqrt{20}$, or $2\sqrt{5}$.
  - Triangles.
    - Pythagorean Theorem: theorem put forth by Pythagoras back in old-school Greece that says that the squares of the legs of a right triangle equal the square of its hypotenuse (longest side).

Remember this only works in right triangles! Equation: $a^2 + b^2 = c^2$

- Triangle Inequality Theorem. Says that the sum of any two sides of a triangle must be greater than the length of the third side. If the problem asks for the greatest possible integer length of a third side, add the two given sides together and subtract 1. If the problem asks for the least possible integer value, subtract the two given sides and add 1.
  - Ex: The three sides of a triangle are 6, 9, and m. What is the product of the least and greatest possible integer values of m? Answer: the greatest m can be is 14 (6+9-1) and the least is 4 (9-6+1). The answer is (14)(4) = **56**.
- Area = ½ base · height {NOTE: Remember that height is the VERTICAL distance from the highest point to the base, meaning it forms a right angle with the base. If you're using a slant height (the length of a slanted side length) as your height, you've gone wrong.}
- The three angles of a triangle add up to 180.

o Properties of Polygons (formulas are in the formula sheet at bottom)
- Parallelogram. Two sets of congruent, parallel sides. Opposite angles are equal. Adjacent angles are supplementary.
- Rhombus. Parallelogram with special features: all four sides are equal and diagonals are perpendicular.
- Rectangle. Parallelogram with special features: all four angles are right.
- Square. Parallelogram with special features: all four sides are equal, all four angles are right, and diagonals are perpendicular.

- Trapezoid. One set of parallel (but non-congruent) sides. That's it! In an isosceles trapezoid the two legs are congruent, and both bottom & top angles are congruent.
- Names of Some Polygons. In increasing number of sides, the polygons are called triangle, quadrilateral, pentagon, hexagon, heptagon, octagon, nonagon, decagon, hendecagon, dodecagon, etc.

- **Radian-Degree Conversion**
  - To say something is "in radians" is simply to say it is "in terms of pi." All we have to know in terms of the test is that $\pi = 180$. Simply substitute one in for the other to convert.
    - Ex: Convert 225° into radians. $\dfrac{225 (\pi)}{180} = 5\pi/4$

    - Ex: Convert $\dfrac{7\pi}{12}$ into degrees. $\dfrac{7(180)}{12} = 105°$.

keep it up!

NOTE: I will only here expand upon those topics I feel require some explanation. Others—those that are neatly summed up by their formulas, such as percent change or standard equation of a circle—will be relegated to the formula pages at the end of this document.

- **Ratios.**
  - What is important in ratio equations is that what the numerator on one side signifies, the other should signify. Ratios are all about consistency.
  - Ex: The ratio boys to girls is 7 to 9. If there are 176 total kids, how many more girls than boys are there?
    - Note here that if we set up the ratio "7boys/9girls" on the left of our equation, we have nothing to add on the right—the question does not give us absolute numbers of either. This inconsistency is what throws people. What we need to do is get a term that matches on both sides, and then we can solve the equation. In this case, that matching term will be "total kids." In the first ratio we can add the terms to find that for every 7 boys there are 16 total kids. And for every 9 girls there are 16 total kids. So my ratio equation will read:

      7boys/16 total = x boys/176 total

      16x = 1232   (we have cross multiplied)

      x = 77 boys... ∴ there are 176 – 77 = 99 girls.

      ∴ there are 22 more girls than boys

- o Ex: The ratio of dimes to nickels is 6 to 5. The ratio of nickels to pennies is 4 to 3. What is the ratio of pennies to dimes?
    - In this case we have a common term (nickels) in two different ratios, but the number of nickels in each is distinct. We need them to match. Therefore, we will multiply each ratio by some number (it can be a different number for each ratio) so that the number of nickels in each is identical. In this problem, I will multiply the first ratio by 4 and the second ratio by 5, giving me the following ratios:

      "There are 24 dimes and 20 nickels... and there 20 nickels and 15 pennies." Since nickels numbers match, we can pair the other two terms into their own ratio. "There are 24 dimes and 15 pennies." Because we paid attention to the question, our answer is 15:24, or **5:8**.

- **Inscribed Figures & Shaded Area.**
    - o Conceptually, all we have to do here is find the area of the larger shape and subtract from it the area of the smaller shape(s) that are inscribed therein. Because these shapes are invariably different geometric figures (a circle in a square, a triangle in a circle, etc), the problem is asking us to know several area formulas. The key conceptual trick is this: notice that certain segments pertain to both figures at once. For example, that triangle leg? It's also a radius of the circle! That diameter of the circle? If draw it straight up and down, you'll notice it's identical to the length of the square! Yay!!! See the quizzes for some examples.

- **Created Symbols**
    - o These tricky devils are just dressed-up functions masquerading as something "weird." It is really simply testing the student's ability not to get overwhelmed by the unfamiliar, and to follow directions.
        - Ex: Let $a \clubsuit b = a^2 - b^2$. What is the value of $-3 \clubsuit 4$?
        $(-3)^2 - (4)^2 \Rightarrow 9 - 16 \Rightarrow$ **-7.**
        - Ex: Let $j \lor k$ represent the number of prime number between j and k, inclusive. How many terms are in the answer set of $6 \lor 39$? OK, let's list out the prime numbers between 6 and 39, inclusive: $\{7, 11, 13, 17, 19, 23, 29, 31, 37\} \Rightarrow$ so $6 \lor 39 =$ **9.**

- **Word problems.**
    - o INVOLVING CONSECUTIVE INTEGERS.
        - Ex: The sum of five consecutive integers is 150. What is the product of the least and greatest numbers in the set?
        We'll assign names to the terms: If n is the first term, the next term is n+1, etc. So:
        $n + (n+1) + (n+2) + (n+3) + (n+4) = 150.$
        $5n + 10 = 150 \Rightarrow n = 28. \therefore$ the numbers are 28, 29, 30, 31, and 32, and
        $28 \cdot 32 =$ **896.**
    - o INVOLVING VENN DIAGRAMS.
        - Ex: In a grade of 128 total kids, there are 80 kids in chorus and 50 in the band. 25 students are in both chorus and band. How many students are in neither one?
        OK, so we'll draw a diagram of overlapping circles. In the overlapping part, we'll write the number 25, as that is the number of kids shared between the two groups. In the left circle (which we'll make the chorus circle), we write the number 55, as

this number represents those kids in ONLY chorus (80 total minus 25 in both). In the right circle, representing the band, we'll write the number 25 to represent the number of kids in ONLY band (50 minus 25). We add the three sub-sections together to find there are a total of 105 students in band and/or chorus. This leaves 128 – 105 = **23 kids that are in neither one.**

- o INVOLVING TOTAL COST.
  - ▪ Ex: The cost of the first minute of a phone call is $2.50 and each additional minute costs 15 cents. If Spencer's phone call costs him $5.65, how long did the call last?

    We'll utilize the formula Total Cost = Fixed Cost + Variable Cost to determine (careful!!!) how many minutes Spencer talks at the 15 cents per minute rate:

    $5.65 = 2.50 + .15x$

    $3.15 = .15x$

    $21 = x$. ∴ Spencer spoke for 21 minutes at the $.15/min rate. We have to add this to the FIRST minute (the one he paid $2.50 for) for a total call time of **22 minutes.**

- ▪ **Special Triangle Rules {all formulas on formula sheet}.**
  - o Special Right Triangles: There are only two "special" right triangles. These triangles are special because their sides are in a constant ratio to each other, which provides a shortcut for calculating side lengths (though the longer process of using Pythagorean Theorem or Law of Sines still works). These are the 45-45-90 ("isosceles") triangle and the 30-60-90 triangle. o SohCahToa: To be used in right triangles only. We take the sine, cosine, or tangent of either of the triangle's acute angles (NOT of the right angle). Once we choose (arbitrarily) which acute angle we'll be using, we carefully choose the trig function that will most effectively produce a result.

o Law of Sines: To be used primarily in non-right triangles, in which the Pythagorean Theorem and SohCahToa are rendered useless. This equation is everyone's favorite: it's a simple matter of ratios. The ratio of the sine of an angle to "its" side (the side opposite that angle) is equal to the same ratio between either of the other pairs in the triangle.

o Law of Cosines: This is the Pythagorean Theorem on steroids. {In fact, the PT is a special case of the Law of Cosines, because in the case of a right triangle the equation produces a final term ending in "cos90°," which equals 0.} To be used in non-right triangles where we are given SAS or SSS.

- **The nth term of an iterative sequence.**

  o A sequence of infinitely repeating terms runs in a pattern. To figure out what the nth term is, no matter how huge, here are our steps: count up how many terms are in the pattern, T; find the closest multiple of T to n; assign T's pattern value to that term; and count from there until you reach n. In layman's terms: the most important term in the pattern is the last one. Keep your eyes on it—and all of its multiples!

    - Ex: Find the 139th term of the sequence 3, 9, -1, 4, 1, 7, 3, 9, -1, 4, 1, 7... OK, so let's count: we have 6 terms in the pattern before it begins repeating. The magic term is the LAST one, remember. Because there are 6 terms in the pattern, we know that every 6th term (i.e., every multiple of 6) will be the term "7." The closest multiple of 6 to 139 is 138, so we know that the 138th term is "7." This means that the next term, the 139th term, is **"3."**

    - Ex: Find the value of i^{75}. OK, so we know that the imaginary number sequence includes a pattern of 4 terms. The fourth term—the important one—is i^4, which equals 1. Therefore, we know that every 4th term is "1." The closest multiple of four to 75 is 72, so the value of the 72nd term is "1." i^{75} is three terms later; its value is **"-i."**

- **"Fishy Questions."**

    o This is a specialized subset of SAT-only math questions in which the question itself is the central clue because it's so atypically specific. In other words, the situation is "fishy." If you can sniff it out, you've cracked the code: you're supposed to work through the problem backwards.

    - Ex: If $a^2 + b^2 = 60$ and $ab = -38$, what is the value of $(a - b)^2$? Notice how fishy the question is! The test has not asked for the value of a, nor the value of b, as it normally would. It has specifically asked for the value of $(a - b)^2$! Suspicious! But we're hip to the game. First step: start with the $(a - b)^2$ term by FOILing it out. It becomes $a^2 - 2ab + b^2$. Notice we can now start plugging in the original clues. We plug in 60 for "$a^2 + b^2$" and -38 for "ab," so that we get $60 - 2(-38) = $ **136.**

- **Counting Principle, Permutations, & Combinations.**

    o The key issue in these problems is figuring out which one of the three to do. On the SAT/ACT, we typically need the Counting Principle or Combinations. CP is used when we're dealing with different categories in combination; Perm/Comb are used when we're dealing with sub-sets within ONE category. Technically speaking, in permutations order DOES matter, whereas in combinations order does NOT matter. The result is that there are always more permutations than there are combinations.

    - Ex: Lisa has 8 pairs of pants, 9 shirts, and 5 pairs of shoes. How many different outfits can she make? This is a Counting Principle question because we're drawing from different categories. $8 \cdot 9 \cdot 5 = $ **360 different outfits.**

- Ex: How many 5-people teams can be made from a group of 9 people? This is a combinations question because we are drawing from within one category and order does NOT matter. Using the formula provided on the formula sheet, we find that the answer is **126 different 5-people team possibilities.**

- Ex: There are 6 chairs in a row. In how many ways can a group of 6 people be seated in those chairs? This is Counting Principle. Draw six slots to represent the six chairs and fill each one with the number of options available to that seat as people go sitting down. This gives you $\underline{6} \cdot \underline{5} \cdot \underline{4} \cdot \underline{3} \cdot \underline{2} \cdot \underline{1}$ = **720 different ways they can be seated.**

- **Circle Sectors.**

    o Sectors are slices of a circle, like a slice of pizza. The question surrounding them is always quite simple: what fraction of the total size of the circle is this sector? Is it one-tenth of the total circle? One fourth? Once we have sorted this out, we know that all the qualities and characteristics of the sector will be in the same proportion to the total circle. For example, if we know that the sector's central angle measures 60°, we know that the sector is one- sixth of the total circle (since 60° is one-sixth of the total of 360°). Therefore we know that the sector's area is one-sixth of the circle's total area, and that the sector's arc length is one- sixth of the total circumference.

- **Graph Translations.**

    o We only need to know a couple of things about basic graph translations. If the original function equation reads $f(x) = x^2$, what have we done to the function's graph when we change the equation to $f(x) = -3(x - 4)^2 + 2$? Well, the original graph is an upward-facing parabola with a vertex of (0, 0). The "3" in front of the x changes the amplitude of the graph, i.e., it makes the graph thinner/steeper. The negative in front of the 3 flips the graph so that it becomes downward-facing. The number inside the parentheses shifts the graph horizontally in the direction opposite its sign, so in this case, "4" shifts the graph to the RIGHT four places. The number all the way to the right shifts the graph vertically in the direction consistent with its sign, so in this case, "2" shifts the graph up two spaces. The resultant parabola has a vertex at (4,2) and is a narrower, downward-facing parabola.

- **Arithmetic & Geometric Sequences.**

  o Simple delineation between the two: arithmetic sequences are those patterns in which the numbers increase or decrease by a fixed amount, or what's called a "common difference." Geometric sequences are those patterns in which the numbers increase or decrease by a fixed multiple, creating what's called a "common ratio." An arithmetic sequence might look like this: 17, 13, 9, 5, 1, -3, -7... (notice the common difference is -4.) A geometric sequence might look like this: 18, 6, 2, 2/3, 2/9, 2/27, 2/81... (notice the common ratio is 1/3).

  - Ex: The terms in a sequence are 8, 2, .5, .125... What is the 6th term in the sequence? To answer this problem, we need first to delineate: is it an arithmetic or a geometric sequence? If it were arithmetic, it would continue to go down by a consistent amount (in this case, by 6, because the first two terms are 8 and 2). Because the third term is not -4, we know the sequence is not arithmetic. When we try geometric instead, we see that we are dividing each term by 4 to get the subsequent term. So it's geometric and the common ratio (r) = ¼. Therefore the 16th term = $8(¼)^{6-1}$ = **1/128**

- **Round Trips.**

  o James drives 20mph to school and then returns at 30mph. The total round trip takes 6 ½ hours. What is James's average rate for the entire round trip? The temptation here is to assume that the "average" rate will simply be the mathematical average of 20 and 30, or 25mph. Wrong! This is a weighted average question, and requires several steps. First, we'll write a separate equation for each leg of the trip. $D = 20t_1$ represents the first leg; $D = 30t_2$ represents the second. Important: notice that in both equations we use simply "D" instead of using subscripts. We do this because we recognize that the distance of each trip is identical. Now we can substitute for D, giving us the equation $20t_1 = 30t_2$. So far, so good. Now we'll add the equation $t_1 + t_2 = 6.5$ to the mix, because we know that the sum of the two trips' time is 6.5. Solving for t_1 in this equation, we get $t_1 = 6.5 - t_2$; we'll now plug in for t_1 into our $20t_1 = 30t_2$ equation. We get: $20(6.5 - t_2) = 30t_2 \Rightarrow t_2 = 2.6$ hours. $\therefore t_1 = 3.9$ hours. We plug these numbers into our original equations and find that the distance from James's home to his school is 78 miles, which means the total round trip is 156 miles. Whew! One more equation: $156 = R_{avg} \cdot 6.5 \Rightarrow R_{avg} =$ **24mph.**

- **Graphing Inequalities.**

  o After shifting terms around so that it reads in the form of y=mx+b, except with an inequality sign in place of the equals sign, graph the line as you typically would. However, as a final step we now plug (0,0) into the inequality as a test point. If this ordered pair makes the inequality "correct" (for example, if after plugging in you get something like -3<5, which is true), then shade the graph on the (0,0) side of the line; if, however, (0,0) makes the inequality "incorrect," shade on the side of the line opposite that point. [NOTE: you can use any test point you like, but (0,0) is considered the easiest.]

- **Matrices.**

  o The tests only ask that the student know what the resulting matrix would be if two original matrices were multiplied, or if a certain number (scalar) were to be multiplied by a matrix. Simple rule: In the latter case, simply distribute the scalar term to every term in the matrix. In the former case, you can only multiply two matrices together if the number of

80

columns in the first matrix equals the number of rows in the second. The resulting product matrix will have the dimensions of {rows of the first}x{columns of the second}. To get its terms, multiply the elements of each ROW of the first matrix by the elements of each COLUMN of the second, and then add the products.

- **Two final notes: Number Lines and "Off by 1."**
  o NUMBER LINES. When given a number line diagram with well-placed number ticks and asked to approximate values of variables placed along it, go ahead and approximate. However, when the problem simply uses words to describe the line ("Points A, B, C, and D lie on a line...") and the problem does NOT say the magic words—"in that order"— then chances are very, very good that your correct answer will be (E) It cannot be determined from the given information. Trust me. o "OFF BY ONE." Off by one questions are misleadingly simple. Be aware that often the obvious, intuitive answer is off from the true answer by one.

    - Ex: Carrie is the 20th person in line, and Dan is 80th person in line. How many people are between them? In this question our intuitive guess is HIGH by one. The answer is not $80 - 20 = 60$, because we don't want to include Carrie. There are **59 people between them**.

    - Ex: Jackie got questions numbered 18 to 63 correct on her biology mid-term. How many questions did she get right in a row? In this question our intuitive guess is LOW by one. The answer is not $63 - 18 = 45$, because we DO want to include the 18th question. Jackie got **46 straight questions right**.

- **Logarithms**
  - o CONVERTING TO/FROM EXPONENTIAL FORM.

    - Ex: Convert $7^2 = 49$ into logarithmic form.

      Remember that $b^x = y \rightarrow \log_b y = x$. So, $7^2 = 49 \rightarrow$ **log₇49=2**

  - o LOG PROPERTIES.

    - Ex: Expand the following: $\log_a \frac{xy^2}{z}$.
      To get this one, we need to know these properties:

      $\log_b xy = \log_b x + \log_b y$

      $\log_b \frac{x}{y} = \log_b x - \log_b y$; $\log_b x^y = y \log_b x$. Therefore, when we expand
      the given logarithm, we get: $\log_a x + 2\log_a y - \log_b z$

- **Imaginary Numbers.**
    - o INVOLVING EXPONENTS

        - Ex: What is the value of i^{427} ?

        To solve this one, we need to remember that there are only four

        possible values for i : $i^0 = 1$, $i^1 = i$, $i^2 = -1$, $i^3 = -i$, $i^4 = 1$... and the pattern repeats. What we really have is a sequence with four distinct terms. So we need to take our exponent (427) and divide it by 4 (representing the four possibilities). The remainder from our division represents our reduced exponents, which gives us the answer.

        $\frac{427}{4} = 106.75$ or $106\ r3$. The remainder of 3 tells us that

        $i^{427} = i^3 = $ -i

    - o INVOLVING FOILING

        - Ex: What is the product: $(5 + 3i)(2i - 7)$?
        We need to foil these two complex binomials:

        $(5 + 3i)(2i - 7) = 10i - 35 + 6i^2 - 21i$. Combine like terms:
        -35-11i+6i²

        Remember that $i^2 = -1 \rightarrow -35 - 11i + 6(-1) = $ **-41-11i**

- **Conic Sections.**
    - o FINDING THE VERTEX FROM STANDARD FORM.
        - Ex: What is the vertex of the following quadratic equation:

        $f(x) = 2x^2 - 8x + 15$?

The given equation is in standard form for a parabola. Note that the vertex of any quadratic equation in standard form can be found using $(\frac{-b}{2a}, f(\frac{-b}{2a}))$.

First we find the x-value of the vertex: $\frac{-b}{2a} = \frac{-(-8)}{2(2)} = 2$. Then we find the y-value of the vertex by plugging in the x-value we found: $f(2) = 2(2)^2 - 8(2) + 15 = 7$

Therefore, the vertex is (2,7).

- o COMPLETING THE SQUARE.
  - Ex: Convert the following quadratic equation into vertex form:

  $f(x) = 2x^2 - 8x + 15$.

  We need to complete the square in order to answer this one. Completing the square basically means that we want to create a perfect square trinomial by adding/subtracting to/from our C-value. Here are the steps you need to follow:

  Factor out A: $f(x) = 2x^2 - 8x + 15 \rightarrow f(x) = 2(x^2 - 4x) + 15$

  Add/Subtract $(\frac{b}{2})^2$ to the binomial and to the constant:

  $(\frac{b}{2})^2 = (\frac{-4}{2})^2 = 4 \rightarrow$

  $f(x) = 2(x^2 - 4x + 4) + 15 - (2^*)(4) = 2(x^2 - 4x + 4) + 7$

  *Note: We multiply by 2 because the trinomial is multiplied by 2. Now, we factor the trinomial: $f(x)=2(x-2)^2+7$ to get our final answer.

Trigonometric Identities

- o USING TRIG IDENTITIES TO SOLVE EQUATIONS

  - Ex: What is the value of $\frac{csc(x)}{cot(x)}$?

    First recall our basic identities: $csc(x) = \frac{1}{sin(x)}$ and $cot(x) = \frac{cos(x)}{sin(x)}$.

    Plug in the identities and solve:

    $$\frac{\frac{1}{sin(x)}}{\frac{cos(x)}{sin(x)}} = \frac{1}{sin(x)} * \frac{sin(x)}{cos(x)} = \frac{1}{cos(x)} = sec(x)$$

- **Unit Circle**

  - o DRAWING/SOLVING TRIANGLES ON UNIT CIRCLE

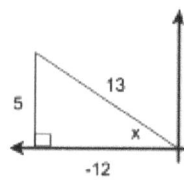

  - Ex: If $sin(x) = \frac{5}{13}$ and $\frac{\pi}{2} < x < \pi$, find $cos(x)$

    We use the information to draw our reference triangle. The fact

    that $\frac{\pi}{2} < x < \pi$ tells us that our triangle is in the second quadrant,

    so we know cosine will be negative. Use the Pythagorean

    theorem (or know your triples!) to solve for the third side of the

    triangle.

    Remember that $cos\ (x) = \frac{adjacent}{hypotenuse}$, so $cos(x) = \frac{-12}{13}$

"PURE MATHEMATICS IS,
IN ITS OWN WAY, THE POETRY
OF LOGICAL IDEAS."

- Albert Einstein

MATH FORMULAS NEEDED FOR BOTH SAT & ACT

Coordinate Geometry

Slope: $m = \dfrac{rise}{run} = \dfrac{y_2 - y_1}{x_2 - x_1}$

Distance between ordered pairs:

$$d = \sqrt{(x_2 - x_1)^2 + (y_2 - y_1)^2}$$

Midpoint: $\left(\dfrac{x_2 + x_1}{2}, \dfrac{y_2 + y_1}{2}\right)$

Equation of a line:

Standard form: $Ax + By = C$

Slope-intercept form: $y = mx + b$

$m = slope$

$b = y - intercept$

Point-slope form: $(y - y_1) = m(x - x_1)$

Laws:

➤ The slopes of parallel lines are equal
➤ The slopes of perpendicular lines are opposite reciprocals

Line Reflections:

$r_{x-axis}\,(x, y) = (x, -y)$
$r_{y-axis}\,(x, y) = (-x, y)$
$r_{y=x}\,(x, y) = (y, x)$
$r_{y=-x}\,(x, y) = (-y, -x)$

Rotations:

$R_{90°}\,(x, y) = (-y, x)$
$R_{180°}\,(x, y) = (-x, -y)$
$R_{270°}\,(x, y) = (y, -x)$
$R_{-90°}\,(x, y) = (y, -x)$

Translation:

$T_{a,b}\,(x, y) = (x + a, y + b)$

Dilation

$D_k\,(x, y) = (kx, ky)$

Triangles

Pythagorean Theorem: $a^2 + b^2 = c^2$

Pythagorean Triples:

$3 - 4 - 5$
$5 - 12 - 13$
$8 - 15 - 17$
$7 - 24 - 25$
$9 - 40 - 41$

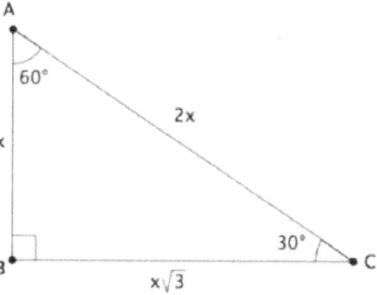

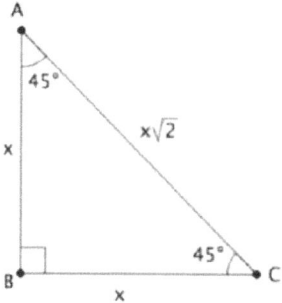

SohCahToa:

$\sin\theta = \dfrac{opposite}{hypotenuse}$

$\cos\theta = \dfrac{adjacent}{hypotenuse}$

$\tan\theta = \dfrac{opposite}{adjacent}$

Definition and Laws:

➤ Any side of a triangle is shorter than the sum of the other two sides (Triangle Inequality Theorem): $a - b < c < a + b$
➤ The sum of the interior angles is $180°$

Special Right Triangles:

Law of sin: $\dfrac{\sin A}{a} = \dfrac{\sin B}{b} = \dfrac{\sin C}{c}$

Law of cosine: $c^2 = a^2 + b^2 - 2ab \cdot \cos C$

Area: $A = \dfrac{1}{2}bh \qquad A = \dfrac{1}{2}ab \cdot \sin C$

Heron's Formula for Area of a Triangle:

$s = \dfrac{a + b + c}{2}$

$A = \sqrt{s(s - a)(s - b)(s - c)}$

BRAINSTORM® TUTORING & ARTS

Circles

Diameter/Radius Conversion: $d = 2r$

Circumference: $C = 2\pi r$ or $C = \pi d$

Area: $A_{circle} = \pi r^2$

Length of Arc/Area of Sector:
$$\frac{central\ \angle}{360°} = \frac{L_{arc}}{C} = \frac{A_{sector}}{A}$$

Other Polygons:

Perimeter:: $P_{square} = 4s \qquad P_{quad} = 2l + 2w$

Area:
$$A_{square} = s^2 = \frac{1}{2}d_1 \cdot d_2$$
$$A_{rhombus} = \frac{1}{2}d_1 \cdot d_2$$
$$A_{parallelogram} = bh$$
$$A_{trapezoid} = \frac{1}{2}(b_1 + b_2)h$$
$$A_{regular\ polygon} = \frac{1}{2}aP$$
$$a = apothem$$
$$P = perimeter$$

Surface Area:
$$SA_{cube} = 6s^2$$
$$SA_{rect\ prism} = 2lw + 2lh + 2wh$$
$$SA_{cylinder} = 2\pi rh + 2\pi r^2$$

Volume:
$$V_{cube} = s^3$$
$$V_{rect\ prism} = lwh$$
$$V_{cylinder} = \pi(r^2)h$$
$$V_{cone} = \frac{1}{3}\pi r^2 h$$
$$V_{pyramid} = \frac{1}{3}Bh \quad (B = Area\ of\ the\ base)$$

Length of diagonal through cube: $d = s\sqrt{3}$

Number of diagonals in a polygon:
$$\#_{diagonals} = \frac{n(n-3)}{2}$$

Interior Angles:
$$S_{interior\ angles} = 180(n - 2)$$
$$m_{interior\ angle} = \frac{180(n-2)}{n}\text{(regular polygons only)}$$

Exterior Angles:
$$S_{exterior\ angles} = 360°$$
$$m_{each\ exterior\ angle} = \frac{360°}{n}$$

Standard Form Equation of a Circle:
$$(x - h)^2 + (y - k)^2 = r^2$$
$$Center: (h, k)$$
$$r = length\ of\ radius$$

Laws:
- A circle is 360°
- $central\ \angle = mArc$
- $inscribed\ \angle = \frac{1}{2}mArc$

Laws: Properties of Select Quadrilaterals

Parallelogram:
- Opposite sides are congruent and parallel
- Opposite angles are congruent
- Consecutive angles are supplementary
- Diagonals bisect each other

Rectangle
- All properties of parallelogram +
- All angles are congruent and measure 90°
- Diagonals are congruent

Rhombus
- All properties of parallelogram +
- All sides are congruent
- Diagonals are perpendicular bisectors

Square
- All properties of parallelogram, rectangle, and rhombus

Trapezoid
- One pair of opposite sides are parallel
- Each lower base angle is supplementary to upper base angle on same side

Names of select polygons:
- 3 sides=triangle
- 4 sides=quadrilateral
- 5 sides=pentagon
- 6 sides=hexagon
- 7 sides=heptagon
- 8 sides= octagon
- 9 sides=nonagon
- 10 sides=decagon

Basic Algebra

Rate: $\quad r = \frac{d}{t} \quad\quad d = rt \quad\quad t = \frac{d}{r}$

Average: $\quad average = \frac{sum\ of\ terms}{number\ of\ terms}$

Cost: $total\ cost = fixed\ cost + variable\ cost$

Percent Change: $\%\ change = \frac{new - original}{original}\ x100$

Direct Proportion: $\frac{x_1}{y_1} = \frac{x_2}{y_2}$

Indirect/Inverse Proportion: $x_1 \cdot y_1 = x_2 \cdot y_2$

Factoring the difference of two squares:
$$a^2 - b^2 = (a + b)(a - b)$$
Factoring Sum/Difference of Cubes:
$$a^3 + b^3 = (a + b)(a^2 - ab + b^2)$$
$$a^3 - b^3 = (a - b)(a^2 + ab + b^2)$$

Exponents and Radicals:

$$x^0 = 1 \qquad\qquad\qquad 9^0 = 1$$
$$x^1 = x \qquad\qquad\qquad 9^1 = 9$$
$$x^2 = x^2 \qquad\qquad\qquad 9^2 = 81$$
$$x^{-2} = \frac{1}{x^2} \qquad\qquad 9^{-2} = \frac{1}{9^2} = \frac{1}{81}$$
$$x^{\frac{1}{2}} = \sqrt{x} \qquad\qquad 9^{\frac{1}{2}} = \sqrt{9} = 3$$
$$x^{\frac{1}{3}} = \sqrt[3]{x} \qquad\qquad 9^{\frac{1}{3}} = \sqrt[3]{9}$$

$$x^a \cdot x^b = x^{a+b}$$
$$\frac{x^a}{x^b} = x^{a-b}$$
$$(x^a)^b = x^{ab}$$

$$\sqrt{x} + \sqrt{y} = \sqrt{x} + \sqrt{y} \qquad\qquad \sqrt{7} + \sqrt{5} = \sqrt{7} + \sqrt{5}$$
$$a\sqrt{x} + b\sqrt{x} = (a + b)\sqrt{x} \qquad \sqrt{7} + 3\sqrt{7} = 4\sqrt{7}$$
$$\sqrt{x} \cdot \sqrt{y} = \sqrt{xy} \qquad\qquad \sqrt{7} \cdot \sqrt{5} = \sqrt{35}$$
$$\frac{\sqrt{x}}{\sqrt{y}} = \sqrt{\frac{x}{y}} \qquad\qquad \frac{\sqrt{27}}{\sqrt{3}} = \sqrt{\frac{27}{3}} = \sqrt{9} = 3$$
$$\sqrt{x^2 y} = x\sqrt{y} \qquad\qquad \sqrt{18} = \sqrt{9} \cdot \sqrt{2} = 3\sqrt{2}$$

Definitions and Laws:
- ➤ Mean is the arithmetic average (found using average formula)
- ➤ Median: middle number when numbers are in numerical order
- ➤ Mode: number that occurs most often
- ➤ Range: difference between lowest and the highest number

Dot Structure: $n = \#\ of\ rows$
$$total\ \#\ of\ dots = \frac{n^2 + n}{2}$$

Advanced Algebra

Combination: $nCr = \frac{n!}{(n-r)!\,r!}$

Arithmetic Sequences/Series:
$$a_n = a_1 + (n - 1)d$$
$$S_n = (a_1 + a_n)\left(\frac{n}{2}\right)$$

Geometric Sequences/Series
$$a_n = a_1 \cdot r^{n-1}$$
$$S_n = a_1 \left(\frac{1 - r^n}{1 - r}\right)$$
$$S_\infty = \frac{a_1}{1 - r}$$

Conic Sections:

Parabola:

Standard Form Equation	$y = ax^2 + bx + c$
Direction of Opening	If $a > 0$, opens up If $a < 0$, opens down
Vertex	$\left(-\dfrac{b}{2a}, f\left(-\dfrac{b}{2a}\right)\right)$
Axis of Symmetry	$x = -\dfrac{b}{2a}$
Y-intercept	$(0, c)$

Quadratic Formula:

$$x = \frac{-b \pm \sqrt{b^2 - 4ac}}{2a}$$

Vertex Form Equation	$y = a(x - h)^2 + k$	$x = a(y - k)^2 + h$
Direction of Opening	If $a > 0$, opens up If $a < 0$, opens down	If $a > 0$, opens right If $a < 0$, opens left
Vertex	(h, k)	(h, k)

Ellipse:

Standard Form Equation	$\dfrac{(x-h)^2}{a^2} + \dfrac{(y-k)^2}{b^2} = 1$	$\dfrac{(x-h)^2}{b^2} + \dfrac{(y-k)^2}{a^2} = 1$
Center	(h, k)	(h, k)

Hyperbola:

	Transverse Axis: Horizontal	Transverse Axis: Vertical
Standard Form Equation	$\dfrac{(x-h)^2}{a^2} - \dfrac{(y-k)^2}{b^2} = 1$	$\dfrac{(x-h)^2}{b^2} - \dfrac{(y-k)^2}{a^2} = 1$
Center	(h, k)	(h, k)

MATH FORMULAS NEEDED FOR THE ACT

Imaginary Numbers:

$$i^0 = 1$$
$$i = \sqrt{-1}$$
$$i^2 = -1$$
$$i^3 = -i$$
$$i^4 = 1$$

Exponential & Logarithmic Forms:

$$y = b^x \leftrightarrow \log_b y = x$$

Properties of Logarithms:

$$\log pq = \log p + \log q$$
$$\log \frac{p}{q} = \log p - \log q$$
$$\log p^q = q \cdot \log p$$

Radian/Degree Conversion: $\pi = 180°$

Trigonometric Identities:

$$\tan x = \frac{\sin x}{\cos x}$$
$$\sec x = \frac{1}{\cos x}$$
$$\csc x = \frac{1}{\sin x}$$
$$\cot x = \frac{1}{\tan x} = \frac{\cos x}{\sin x}$$

$$\sin^2 x + \cos^2 x = 1$$
$$1 + \tan^2 x = \sec^2 x$$
$$1 + \cot^2 x = \csc^2 x$$

$$\sin 2x = 2 \sin x \cos x$$
$$\cos 2x = \cos^2 x - \sin^2 x$$
$$= 2\cos^2 x - 1$$
$$= 1 - 2\sin^2 x$$

Unit Circle:

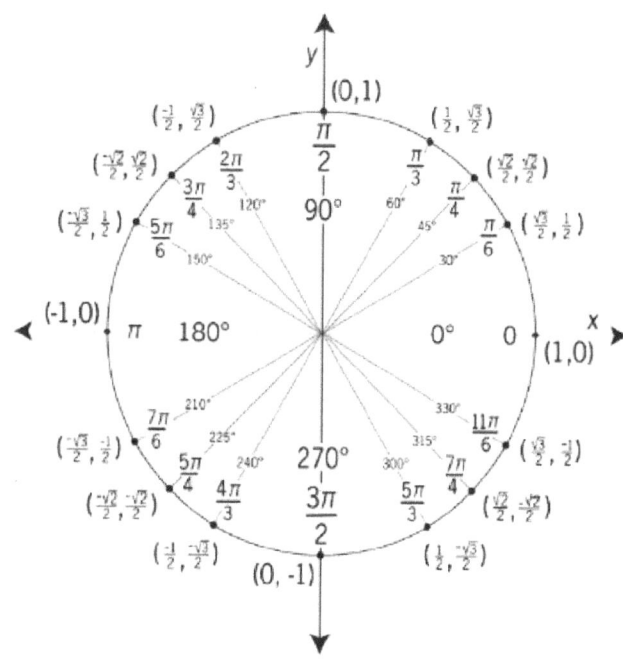

Matrix Algebra

$$A = \begin{bmatrix} a & b \\ c & d \end{bmatrix}$$

$$kA = \begin{bmatrix} ka & kb \\ kc & kd \end{bmatrix}$$

$$\begin{bmatrix} a & b \\ c & d \end{bmatrix} \pm \begin{bmatrix} w & x \\ y & z \end{bmatrix} = \begin{bmatrix} a \pm w & b \pm x \\ c \pm y & d \pm z \end{bmatrix}$$

Determinant of a 2x2 Matrix:

$$A = \begin{bmatrix} a & b \\ c & d \end{bmatrix} \to |A| = ad - cb$$

Matrix Multiplication

$$\begin{bmatrix} a & b \\ c & d \end{bmatrix}\begin{bmatrix} w & x \\ y & z \end{bmatrix} = \begin{bmatrix} aw + by & ax + bz \\ cw + dy & cx + dz \end{bmatrix}$$

$$\begin{bmatrix} a & b & c \\ d & e & f \\ g & h & i \end{bmatrix}\begin{bmatrix} q & r & s \\ t & v & w \\ x & y & z \end{bmatrix} = \begin{bmatrix} aq + bt + cx & ar + bv + cy & as + bw + cz \\ dq + et + fx & dr + ev + fy & ds + ew + fz \\ gq + ht + ix & gr + hv + iy & gs + hw + iz \end{bmatrix}$$

PERFECT SQUARES	PERFECT CUBES	PRIME NUMBERS
$1^2 = 1$	$1^3 = 1$	2
$2^2 = 4$	$2^3 = 8$	3
$3^2 = 9$	$3^3 = 27$	5
$4^2 = 16$	$4^3 = 64$	7
$5^2 = 25$	$5^3 = 125$	11
$6^2 = 36$	$6^3 = 216$	13
$7^2 = 49$	$7^3 = 343$	17
$8^2 = 64$	$8^3 = 512$	19
$9^2 = 81$	$9^3 = 729$	23
$10^2 = 100$	$10^3 = 1000$	29
$11^2 = 121$		31
$12^2 = 144$		37
$13^2 = 169$		41
$14^2 = 196$		43
$15^2 = 225$		47
$16^2 = 256$		53
$17^2 = 289$		59
$18^2 = 324$		61
$19^2 = 361$		67
$20^2 = 400$		71
$21^2 = 441$		73
$22^2 = 484$		79
$23^2 = 529$		83
$24^2 = 576$		89
$25^2 = 625$		97

HERE WE GO...TIME TO TEST YOUR SKILLS.
YOU GOT THIS!

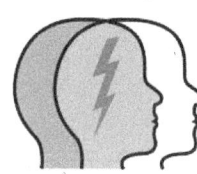

Directions: Answer the following questions by showing all your work.

Note that figures are **not** drawn to scale.

Be sure to use correct units in your answers.

(1) The diameter of a circle is 8cm. What is its circumference? 8π What is its area? 16π

↙ radius is 4 $2\pi 4 = 8\pi$ $a = \pi 4^2$

(2) How many diagonals are there in a 24-gon? 252 What is the sum of its interior angles? $3,960$

Its exterior angles? $360°$

↙ always $360°$

$\dfrac{n(n-3)}{2}$ $\dfrac{24(24-3)}{2}$ $\dfrac{504}{2}$ $(n-2)180°$

$(24-2)180 = 3,960$

(3) What is the sum of x and y? _____

$180 - 84 - 5 - 3x + 6x - 8$ 6x-8

$180 - 96$

$y = 84$

5-3x   y   96°

(4) Given $c \parallel d$, what is $m\angle Q$? $147°$

parallel

$180°$

(5) In right $\triangle JQL$, $\cos J = \dfrac{18}{82}$. What is $\tan J$? $\dfrac{80}{18}$ $\sec J$? $\dfrac{82}{18} \rightarrow \dfrac{42}{9}$

$\cos = \dfrac{adj}{hyp}$

$\sec = \dfrac{hyp}{adj}$ $\dfrac{82}{18}$

Soh cah toa

$\tan = \dfrac{opposite}{adj}$

$324 + b^2 = 6724$
$-324 \quad -324$

$\sqrt{b^2} = \sqrt{6400}$

$b = 80$

$a^2 + b^2 = c^2$

$18^2 + b^2 = 82^2$

$324 + b^2 = 82^2$

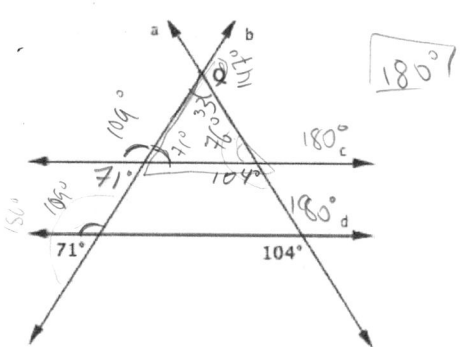

$\tan = \dfrac{80}{18}$

$b = 80$

95

(6) What is the area of a parallelogram with corners A (-1,1), B (3,3), C (8,3), and D (4,1)? _____

(7) The sides of $\triangle MRS$ measure 17, 24, and c. What is the product of the least and greatest possible integer values of c? _____

(8) What is the area of this figure? _____

What is its perimeter? _____

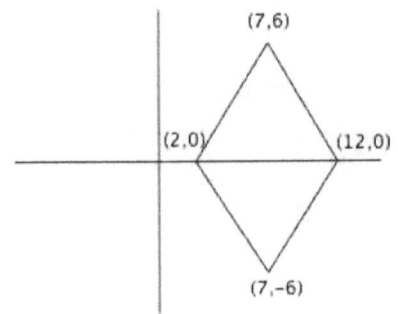

(9) What is the area of $\triangle PAX$? _____

What is its perimeter? _____

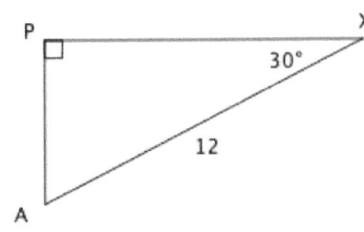

(1) We know that a circle's diameter is the segment that runs from one edge to the other and through the center of a circle. Therefore, we know that the diameter is twice the circle's radius: $d = 2r$ (r=radius). We are given that the diameter is 8cm long. So, $d = 8 = 2r \rightarrow r = 4cm$. The circumference C of a circle is given by the formula: $C_O = \pi d$, where d represents the diameter. Substitute 8 for d and: $C_O = 8\pi$ cm. Now, the area, A of a circle is given by the formula $A_O = \pi r^2$. Substitute 4 for r: $A_O = 16\pi$ cm^2

(2) To find the number of diagonals of a polygon with n number of sides (*note-this can also be the number of angles) we use the formula: $D = \left(\dfrac{n}{2}\right)(n - 3)$, where D represents the number of diagonals. Therefore, in a 24-gon: there are: $D_{24} = \left(\dfrac{24}{2}\right)(24 - 3) = 252$. Now, to find the sum of the interior angles of a polygon with n number of sides, use the formula; $S_{int} = 180(n - 2)$ where S_{int} represents the sum of the polygon's interior angles. Therefore, the sum of the interior angles of a 24-gon is: $S_{int\,24-gon} = 180(24 - 2) = 3960°$. Good to know: the exterior angles of a *any* polygon add up to $360°$. You can prove it to yourself- using polygons like triangles, squares, and hexagons.

(3) Since the sum of the angles of a triangle is $180°$, we can conclude that $6x - 8 + 5 - 3x + y = 180°$. Due to the fact that the angles with measures y and $96°$ are *linear pairs* (supplementary angles), $y + 96° = 180°$. Solving for y in the second equation we have: $y = 84°$. Now, let's substitute this value in the first equation to solve for x: $6x - 8 + 5 - 3x + 84° = 180° \rightarrow 3x + 81 = 180 \rightarrow 3x = 99 \rightarrow x = 33°$. So the sum of x and y is: $x + y = 33° + 84° = 117°$

(4) In the diagram on the right $m\angle b = 71°$, because it is a vertical angle to

the angle with measure $71°$. Also, $\angle c$ and the angle with measure $104°$

are linear pairs (supplementary). Therefore,

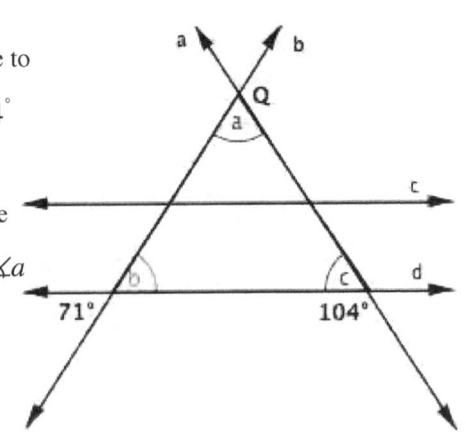

$m\angle c + 104° = 180° \rightarrow m\angle c = 76°$. Because three angles in a triangle

add up to $180°$, $m\angle a + 71 + 76 = 180° \rightarrow m\angle a = 33°$. Notice that $\angle a$

and $\angle Q$ are linear pairs as well. So we conclude that:

$m\angle Q + 33° = 180° \rightarrow m\angle Q = 147°$

(5) In right $\triangle LQJ$ on the right, the cosine of $\angle J$ is the ratio of the angle's

adjacent side to the triangle's hypotenuse: $\cos J = \dfrac{18}{82}$. The tangent of $\angle J$ is the

ratio of the angle's opposite side to its adjacent side: $\tan J = \dfrac{opp}{adj} = \dfrac{?}{18}$ We find

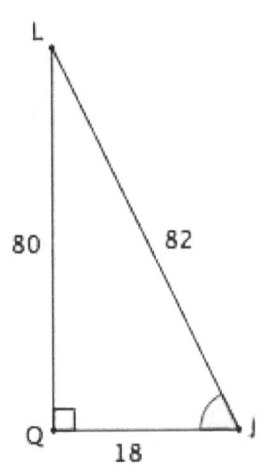

the measure of the opposite side by using the Pythagorean theorem:

$opp^2 + 18^2 = 82^2 \rightarrow opp = \sqrt{82^2 - 18^2} \rightarrow opp = 80$. Therefore,

$\tan J = \dfrac{80}{18} \rightarrow \dfrac{40}{9}$. Now, the secant of $\angle J$ is equal the reciprocal of the cosine

of $\angle J$: $\sec J = \dfrac{1}{\cos J} = \dfrac{82}{18} \rightarrow \sec J = \dfrac{82}{18} \rightarrow. \dfrac{41}{9}$

(6) The area of a parallelogram $\square DABC$ is given by

the formula: $A_\square = (base)(height)$. For this

parallelogram we'll call the base, $\overline{AB} = 5$. What about

the height? Well, the height of this figure is the parallelogram's altitude, or the length of the shortest line

segment between $\overline{DC}$ and $\overline{AB}$ Draw the altitude, $\overline{DH}$. The length of $\overline{DH}$ is the vertical height (y-

coordinate) of point D minus the vertical height (y-coordinate) of point H. Because $D(3,3)$, D's y-

coordinate is 3. But what about H's y-coordinate? Well, point H is on the same line as points A and B,

right? Both A and B have y-coordinates that equal 1. Therefore, H's y-coordinate is 1. So,

$\overline{DH} = y_C - y_H = 3 - 1 \rightarrow \overline{DH} = 2$. Now, let's find area: $A_{\square DABC} = (\overline{AB})(\overline{DH}) = (5)(2) = 10$.

(7) An important rule to know about the lengths of the sides of a triangle: the sum of the lengths of any two sides will be greater than the length of the third side. Therefore, in $\triangle MRS$ with side lengths of 17, 24, and c, we know that: $17 + 24 > c \rightarrow c < 41$. So, the greatest possible integer that c could be is 40. We also know that: $17 + c > 24 \rightarrow c > 7$. So, the smallest possible integer that c could be is 8. Now, we'll find the product of these values: $c_{smallest} \times c_{greatest} = 8 \times 40 = 320$.

(8) The area of a rhombus is given by the formula: $A_{rb} = \frac{1}{2}d_1 d_2$. This rhombus has coordinate points $A(2,0)$, $B(7,6)$, $C(12,0)$ and $D(7,-6)$ with diagonals $\overline{AC}$ and $\overline{BD}$. Draw them. We notice that we can find the measure of $\overline{AC}$ by calculating the positive difference between the x-coordinates of points A and C: $\overline{AC} = x_C - x_A = 12 - 2 \rightarrow \overline{AC} = 10$. Now, to find $\overline{BD}$ we calculate the positive difference of the y-coordinates of points B and D: $\overline{BD} = y_B - y_D = 6 - (-6) \rightarrow \overline{BD} = 12$. So, $d_1 = \overline{AC} = 10$ and $d_2 = \overline{BD} = 12$. Let's substitute these values into out original formula for area of a rhombus: $A_{rb} = \frac{1}{2}(\overline{AC})(\overline{BD}) = \frac{1}{2}(10)(12) \rightarrow A_{rb} = 60$. Now, the perimeter of rhombus ABCD is simply the sum of its sides. Since all the sides of a rhombus are equivalent in length, we have to find only one of the side lengths. We use the distance formula to find the length of $\overline{AB}$:

$$d = \sqrt{(x_2 - x_1)^2 + (y_2 - y_1)^2} \rightarrow d_{\overline{AB}} = \sqrt{(7-2)^2 + (6-0)^2} \rightarrow d_{\overline{AB}} = \sqrt{61}. \text{ So,}$$
$$P_{rb} = \overline{AB} + \overline{BC} + \overline{CD} + \overline{DA} = 4\sqrt{61}$$

(9) Let's first point out that $\triangle PAX$ is a 30-60-90 triangle (we know this because if two angles of a triangle are given to be $30°$ and $90°$, the third must be $60°$) Remember that the ratio of the sides of a 30-60-90 triangle is $x : x\sqrt{3} : 2x$. Let's solve for x: since 12, the hypotenuse, is opposite the $90°$ angle, $2x = 12 \rightarrow x = 6$. Therefore, $\overline{PA} = 6$ and $\overline{PX} = 6\sqrt{3}$ The area of a triangle is given by the formula: $A_\triangle = \frac{1}{2}(base)(height)$. Let's plug our base and height into the equation to find our area:

$A_\triangle = \frac{1}{2}(\overline{PA})(\overline{PX}) = \frac{1}{2}(6)(6\sqrt{3}) \rightarrow A_\triangle = 18\sqrt{3}$. The perimeter is simply the sum of the lengths of the sides of the triangle: $P_{\triangle PAX} = PA + PX + AX = 12 + 6 + 6\sqrt{3} = 18 + 6\sqrt{3}$.

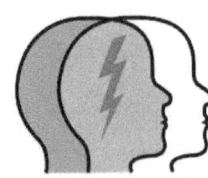

Directions: Answer the following questions by showing all your work.

Note that figures are **not** drawn to scale.

Be sure to use correct units in your answers.

(1) What is the measure of one interior angle in a regular 20-gon? _____

(2) How many diagonals does a nonagon have? _____

(3) If the volume of a cube is 216cm³, what is its surface area? _____ What is the length of the diagonal

running through the center of this cube? _____

all ∡ = 360

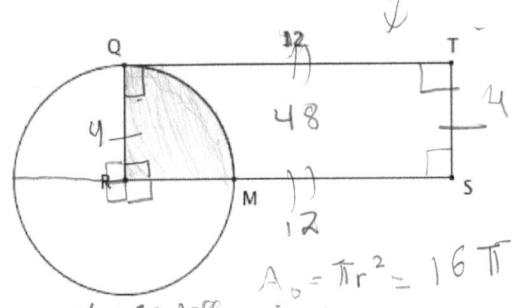

(4) R is the center of the circle and $QRST$ is a rectangle

with area of 48. What is the area of sector QRM ?

$\frac{1}{4} A_6 = $ shaded area $= 4\pi$

$A_\square = lw$

$48 = lw$

$A_0 = \pi r^2 = 16\pi$

same slope, diff y-int

(5) What is the equation of the line passing through $(-1,1)$ and parallel to $4x - 7y = 42$? _____

1) find slope → $y = mx + b$ — $y = \frac{4}{7}x - 6$

new line: $y = \frac{4}{7}x + b = \frac{4}{7}x + \frac{11}{7}$

$1 = \frac{4}{7}(-1) + b$ $b = \frac{11}{7}$

(6) A semicircle with center Q is attached to square

$PONM$. If $\overline{MO} = 16\sqrt{2}$, what is the length of arc

$\overparen{MN}$? _____

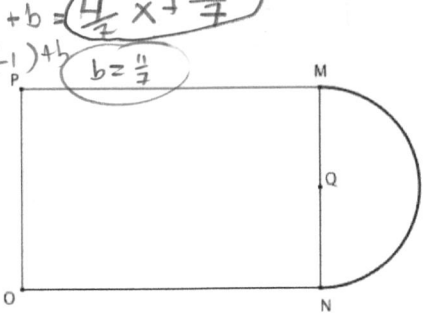

(7) What is the value of $5x - 3$? _____

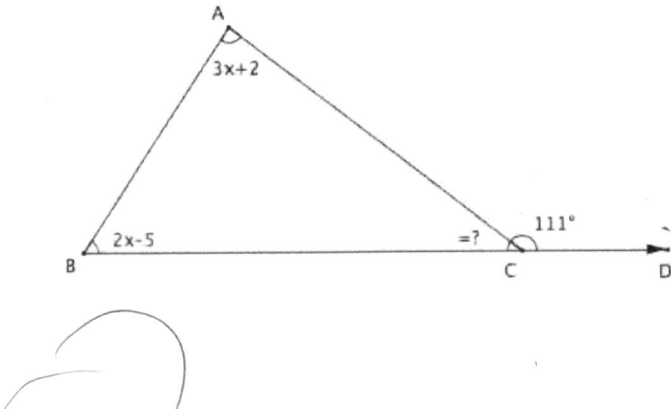

100

(8) Given $a \parallel b$, find the value of y. _____

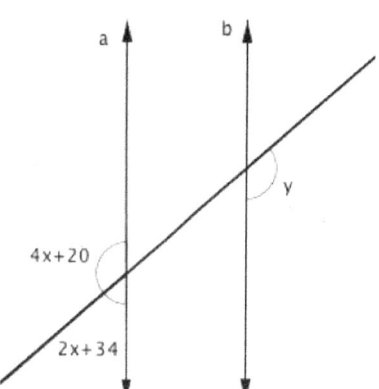

(1) The measure of a single interior angle in a regular polygon with 'n' number of sides is given by the

formula: $m\angle\text{int} = \dfrac{180(n-2)}{n}$. Therefore, for a 20-gon, n=20: $m\angle\text{int} = \dfrac{180(20-2)}{20} \rightarrow m\angle\text{int} = 162°$

.

(2) The number of diagonals in a polygon with 'n' number of sides is given by the formula: (D = number of

diagonals): $D = \left(\dfrac{n}{2}\right)(n-3)$. A nonagon has 9 sides so: $D_{nonagon} = \left(\dfrac{9}{2}\right)(9-3) = 27$.

(3) Let's start with the formula for the volume of a cube: $V_{cube} = s^3$ where 's' is the side length of the cube. We

are given that our cube's volume is: $216cm^3$. Let's plug this in and solve for s:

$V_{cube} = 216 = s^3 \rightarrow s = \sqrt[3]{216} \rightarrow s = 6$ Now, formula for the surface area of a cube is given by:

$SA_{cube} = 6(s^2)$. In this case, $s = 6$. So, $SA_{cube} = 6(6^2) \rightarrow SA_{cube} = 216cm^2$. Finally, the length of the

diagonal that runs through the center of a cube is given by: $d_{cube} = s\sqrt{3}$. Since $s = 6$, $d_{cube} = 6\sqrt{3}$.

(4) Let's begin by finding the measure of the side of the rectangle,

$\overline{QR}$. The area of a rectangle is given by the formula:

$A_{□} = (base)(height)$. In this case,

$A_{rect} = 48 = (12)(\overline{QR}) \rightarrow \overline{QR} = 4$. Notice that $\overline{QR} = r$, the

radius of circle R. So, the area of the circle R, (given by the

formula, $A_○ = \pi r^2$) is $A_{○R} = 16\pi$. But how do we find the area

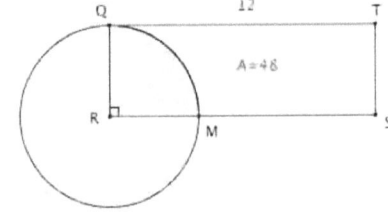

of sector QRM? Well, notice that sector QRM includes a $90°$ angle. Therefore, this sector is $\dfrac{90°}{360°} = \dfrac{1}{4}th$

of the entire circle. This tells us that $A_{\sec QRM} = \dfrac{1}{4}A_○ \rightarrow A_{\sec QRM} = \dfrac{1}{4}(16\pi) = 4\pi$.

(5) We know that parallel lines have the **same slope.** So, we know that any line parallel to $4x - 7y = 42$ will have

the same slope. Let's find the slope of our given line, l_1 by converting its equation into slope-intercept form:

$y = mx + b$ where m=slope. So, $4x - 7y = 42 \rightarrow y = \frac{4}{7}x - 6$. In this case, $m = \frac{4}{7}$. Therefore, in our

parallel line, l_2, the slope, $m = \frac{4}{7}$. So far we have the equation for l_2 as: $y = \frac{4}{7}x + b$. We are given that l_2

passes through the point $(-1, 1)$. We plug in the x and y- coordinates of this point into the equation to find the

value of b: $1 = \frac{4}{7}(-1) + b \rightarrow b = 1\frac{1}{7}$. Therefore, the equation of our parallel line l_2: $y = \frac{4}{7}x + 1\frac{1}{7}$

(6) Let's begin by drawing $\overline{MO}$ the diagonal of square $PONM$ and the hypotenuse of $\triangle MON$. Notice that this triangle is an isosceles right triangle (Right triangles with legs of equal length). The ratio of the sides of an isosceles right triangle is $x : x : x\sqrt{2}$. Therefore,

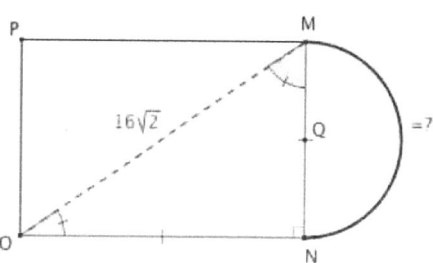

$\overline{MO} = x\sqrt{2} = 16\sqrt{2} \rightarrow x = 16$. So, $\overline{MN} = x = 16$. But, $\overline{MN}$ is not only the side of the square $PONM$ but also the diameter of attached semicircle Q. The circumference of a semicircle is given by the formula $C_{semi \odot Q} = \frac{1}{2}\pi d \rightarrow \overparen{MN} = C_{semi \odot Q} = 8\pi$.

(7) We know that $\angle ACB$ and $\angle ACD$ are linear pairs. Therefore, $m\angle ACB + m\angle ACD = 180°$. Solve to find $m\angle ACB$: $m\angle ACB + 111° = 180° \rightarrow m\angle ACB = 69°$. Since the sum of the angles of a triangle is $180°$: $69 + 3x + 2 + 2x - 5 = 180 \rightarrow 5x = 114$. Therefore, $5x - 3 = 114° - 3 \rightarrow 5x - 3 = 111°$.

(8) Because the angles that are represented by the values $4x + 20$ and $2x + 34$ are linear pairs we know that:

$4x + 20 + 2x + 34 = 180° \rightarrow 6x = 126° \rightarrow x = 21°$. So,

$m\angle PRT = 4(21) + 20 = 104°$. Since, $a \parallel b$ $m\angle PRT = y$

because they represent **alternate exterior angles.** Therefore,

$y = 104°$.

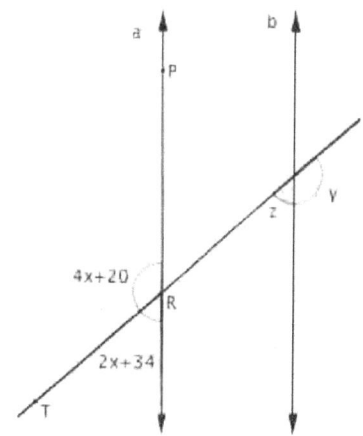

103

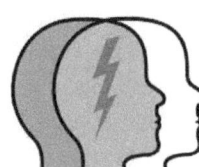

Directions: Answer the following questions by showing all your work.

Note that figures are **not** drawn to scale.

Be sure to use correct units in your answers.

(1) What is the sum of the interior angles of a dodecagon? _____

(2) Given $b \parallel c$, what is the value of z? _____

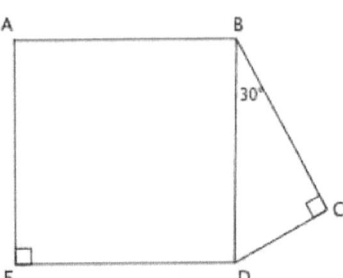

(3) The area of square $AEDB$ is 72 in². What is the

perimeter of $AEDCB$? _____

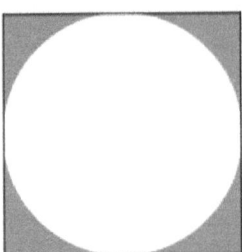

(4) The area of the circle to the right is 18π. What is the area

of the shaded region? _____

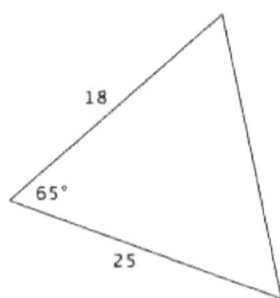

(5) Find the values of the triangle's third side and remaining two

angles. (s)_____ ($m\angle 1$)_____ ($m\angle 2$)_____

(6) What is the surface area of a rectangular prism with dimensions 8cm x 5cm x ½cm? _____

(7) How many diagonals are in an octagon? _____

(8) A cube has a volume of 343m³· What is the length of the diagonal running through the center of the cube? _____

(9) Given J and M are the centers of the cylinder's bases. $\overline{MQ}$ = 12cm. The cylinder's volume is 720π cm³. What is the length of $\overline{JQ}$? _____

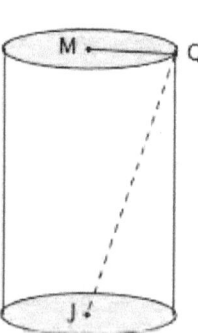

(10) $JAQR$ is a rhombus. What is its area? _____

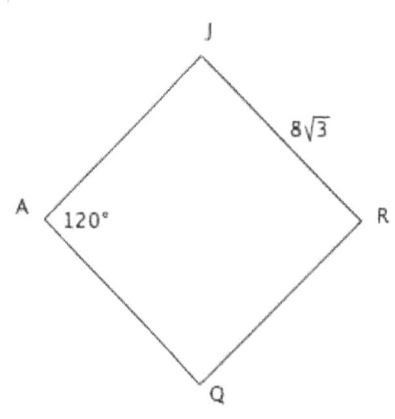

(11) Given $\overline{NO} \parallel \overline{PQ}$, $\overline{MN} = 8$, and $\overline{NO}\big/_{PQ} = {}^{2}\!/_{5}$, what

is the length of $\overline{NP}$? _____

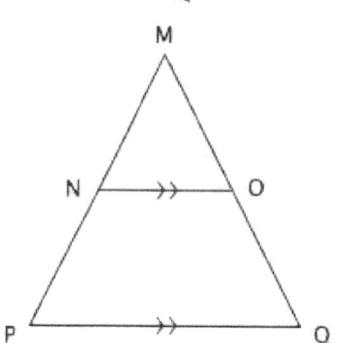

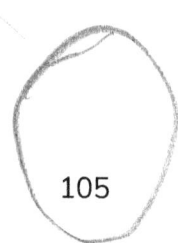

105

(1) We can find the sum of the interior angles of a dodecagon by applying the appropriate formula: The sum of the interior angles of a polygon with n number of sides is $180°(n-2)$. Since there are 12 sides in a dodecagon, $S_{int \angle's} = 180°(12-2) = 1800°$. So, $S_{int \angle's} = 1800°$

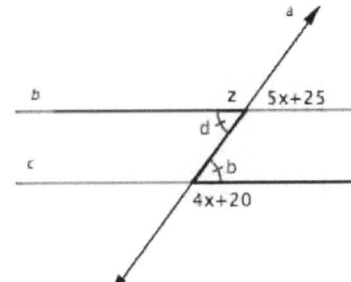

(2) Since $b \parallel c$, $m\angle b = m\angle d$ because $\angle b$ and $\angle d$ are alternate interior angles. Since $\angle b$ and the angle with value $4x+20$ form a *linear pair* $m\angle b + 4x + 20 = 180°$. We also know that $5x + 25 = m\angle d$ because $\angle d$ and the angle with value $5x + 25$ are vertical angles. Substitute the appropriate values and solve for x: $5x + 25 + 4x + 20 = 180°$. Solving, we conclude that $x = 15°$. Now, we must solve for z: z and $5x + 25$ are values of angles that form a linear pair. So, $5x + 25 + z = 180° \rightarrow z = 180° - 5x - 25$. Now, substitute for x ($x = 15°$), and simplify: $z = 180° - 5(15°) - 25 \rightarrow z = 80°$

(3) The area of a square, with side s is given by the formula: $A = s^2$. We are given the area of square $AEDB$ is 72 in². Therefore, $A_{AEDB} = s^2 = 72 \rightarrow s = 6\sqrt{2}$. To find the perimeter of this figure, we have to find the values of $\overline{CD}$ and $\overline{BC}$. We notice that $\triangle BDC$ is a 30-60-90 triangle. We know that the hypotenuse, $\overline{BD} = 6\sqrt{2}$ because it is also a side of the square. The ratio of lengths of the sides of a 30-60-90 triangle is $x : x\sqrt{3} : 2x$ We can solve for x and the remaining sides of the triangle using this ratio. Since the hypotenuse is opposite the largest angle, $\overline{BD} = 2x = 6\sqrt{2} \rightarrow \overline{CD} = x = 3\sqrt{2}$. Therefore, $\overline{BC} = x\sqrt{3} = 3\sqrt{6}$. Now, we can calculate the perimeter: $P_{figure} = \overline{DE} + \overline{EA} + \overline{AB} + \overline{BC} + \overline{CD} \rightarrow P_{figure} = 3(6\sqrt{2}) + 3\sqrt{2} + 3\sqrt{6} \rightarrow P_{figure} = 21\sqrt{2} + 3\sqrt{6}$

(4) The area of a circle, with radius, r, is given by the formula: $A_O = \pi r^2$. Since we are given that the area of the circle is 18π, we can solve for its radius, r: $A_O = 18\pi \rightarrow r = 3\sqrt{2}$ The diameter is twice the radius so, $d = 6\sqrt{2}$. Yet, if we draw the diameter of the circle (on the diagram), we notice that it is equivalent in length to the side of the square. So, $s = 6\sqrt{2}$ and $A_\square = s^2 = \left(6\sqrt{2}\right)^2 = 72$. Now, the area of the shaded region is the area of the square minus the area of the circle: $A_{shadedregion} = A_\square - A_O = 72 - 18\pi = 15.451$.

(5) We are given two sides and an angle. Label the $\triangle ABC$ where $m\angle B = 65°$,

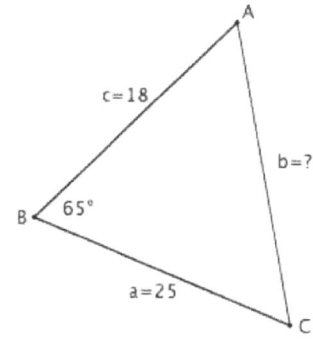

$\overline{AB} = c = 18$ and $\overline{CB} = a = 25$. Let's use the **Law of Cosines** to find the value of b, our third side. The Law of Cosines tells us that

$b^2 = c^2 + a^2 + 2ca\cos B$. Let's substitute in the information we already have and simplify in order to solve for side, b:

$b^2 = 18^2 + 25^2 - 2(18)(25)\cos 65° \rightarrow b = 23.85$. To find the measures of the

other two angles, use the **Law of Sines:** $\dfrac{b}{\sin B} = \dfrac{a}{\sin A} \rightarrow \dfrac{23.85}{\sin 65°} = \dfrac{25}{\sin A}$.

Using cross multiplication and the inverse sine function, solve for A:

$\sin A = \dfrac{25\sin 65°}{23.85} \rightarrow \sin^{-1}(.95) \rightarrow m\angle A = 75.84°$ Now, we find the last angle:

$75.84° + 65° + m\angle C = 180° \rightarrow m\angle C = 39.16°$.

(6) A rectangular prism has 6 faces- 4 sides and 2 bases. Draw one. Notice that each face is a rectangle- and opposite faces are the **same**. The surface area of this prism is the sum of the areas of all the faces. We are given dimensions of our prism. $8 \times 5 \times \frac{1}{2} = length \times width \times height$. So,

$SA_{prism} = 2Face_1 + 2Face_2 + 2Base = 2wh + 2hl + 2wl$. Now, substitute your values and simplify: .

$SA_{prism} = 2\left(\frac{1}{2}\right)(5) + 2\left(\frac{1}{2}\right)(8) + 2(8)(5) \rightarrow SA_{prism} = 93cm^2$

(7) We find the number of diagonals, d in a polygon with n number of sides with the formula $d = \left(\frac{n}{2}\right)(n-3)$
Since $n = 8$ in an octagon, $d = \left(\frac{8}{2}\right)(8-3) \rightarrow d = 20$

(8) A cube's volume is given by the formula: $V_{cube} = s^3$ where s is a side of the cube. We are given that the volume of this cube is 343cm^3. So, $V_{cube} = s^3 = 343 \rightarrow s = \sqrt[3]{343} \rightarrow s = 7$. Now, draw the diagonal of the base of the cube (from the bottom back right corner to the bottom front left corner). Call this diagonal d_b which forms a 45-45-90 triangle with the sides of the cube. Using the ratio of the sides of a 45-45-90 triangle, $d_b = 7\sqrt{2}$. Now, draw the diagonal that runs through the center of the cube (top back right corner to bottom front left corner). Call it d_c. Now, you have created a right triangle in the center of the cube with sides s, d_c and d_b. Use Pythagorean theorem to find.

$$d_c \quad d_c^2 = d_b^2 + s^2 \rightarrow d_c = \sqrt{\left(7\sqrt{2}\right)^2 + 7^2} \rightarrow d_c = 7\sqrt{3}$$

(9) Let's start with finding some missing values: First of all, we know that

$V_{cylinder} = \pi r^2 h$ And, we are given that $V_{cylinder} = 720\pi$ and $r = 12$. Substitute these

values into the equation for the volume of a cylinder and solve for its height, h:

$720\pi = \pi(12)^2 h \rightarrow h = 5$. Draw segment $\overline{JM}$. You can see that this segment is

your height! In other words, $\overline{JM} = h = 5$. Our new triangle, $\triangle MJQ$ is a right triangle

with side ratios 5 -12 – 13 (*you should have these side lengths of a right triangle

memorized like 3- 4- 5)! So, $\overline{JQ} = 13$.

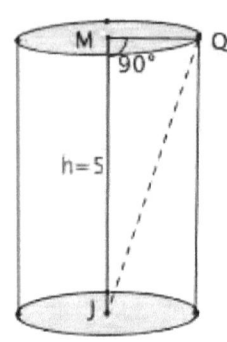

(10) We want to use the formula for the area of the rhombus is: $A_{rhs} = \frac{1}{2} d_1 d_2$. Rhombi are very special

parallelograms. The diagonals of a rhombus meet at perpendicular angles while they bisect each other and the corner

angles. Draw the diagonals of this rhombus. Allow P to be the center intersection point of the diagonals. Focus on

$\triangle JPR$, a 30-60-90 triangle. Using the ratios of the sides of these special triangles, $x : x\sqrt{3} : 2x$ we have

$\overline{JR} = 8\sqrt{3} = 2x \rightarrow x = 4\sqrt{3} = \overline{PR}$, and $\overline{PJ} = x\sqrt{3} = 12$. Again, the diagonals of a rhombus bisect each other so,

$d_1 = 2(12) = 24$ and $d_2 = 2(4\sqrt{3}) = 8\sqrt{3}$. Great!-now let's plug –in these values into the original equation and

simplify: $A_{rhs} = \frac{1}{2}(24)(8\sqrt{3}) \rightarrow A_{rhs} = 166.28$

(11) Since $\overline{NO} \parallel \overline{PQ}$, we can conclude that $\angle MNO \cong \angle MPQ(A)$ and $\angle MQN \cong \angle MQP(A)$ because they are corresponding angles. Also, $\angle M \cong \angle M$. Therefore, $\triangle MPQ$ and $\triangle MNO$ are similar triangles through AAA (angle, angle, angle) similarity ($\triangle MPQ \sim \triangle MNO$). This tells us that sides of each of these triangles are proportionate to each other by a certain ratio. Because we are given that $\overline{NO}/\overline{PQ} = 2/5$, the ratio of the sides of $\triangle MNO$ to those of $\triangle MPQ$ is $2/5$. Therefore, $\overline{MN}/\overline{MP} = 2/5$. So, we set up a proportion and solve for $\overline{MP}$:

$\overline{MN}/\overline{MP} = 2/5 \rightarrow 8/\overline{MP} = 2/5 \rightarrow \overline{MP} = 20$. Now, we solve for $\overline{NP}$: $\overline{NP} = \overline{MP} - \overline{MN} = 20 - 8 \rightarrow \overline{NP} = 12$

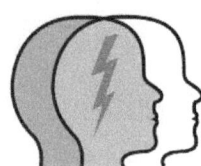

Directions: Answer the following questions by showing all your work.
Note that figures are **not** drawn to scale.
Be sure to use correct units in your answers.

(1) Q is the center of a circle with $\overline{AP} = 6$ cm. What is the length of

$\overset{\frown}{AP}$? _____

What is the area of the shaded region? _____

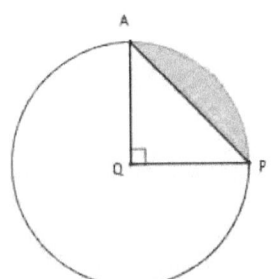

(2) Find the area of the trapezoid. _____

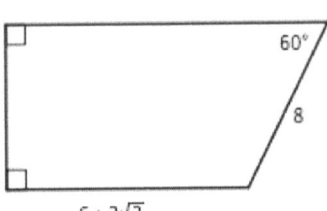

(3) C is the center of a circle with radius $\sqrt{7}$ in. Circle C is inscribed in a quadrilateral. What is the area of the shaded region? _____

What is the perimeter of the shaded region? _____

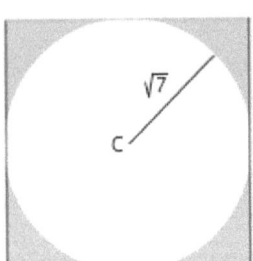

(4) Given the intersecting lines on the right, what is the value (*in degrees*) of $b^2 - d$? _____

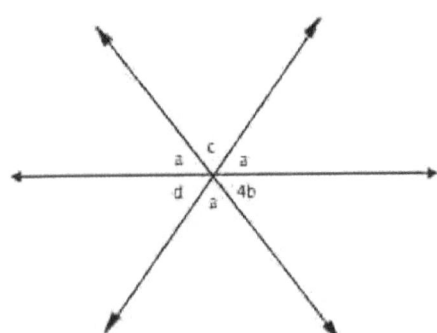

(5) In $\triangle GHX$, $\overline{GH} = 13$, $\overline{HX} = 17$, and $\overline{GX} = 8$. What is the sum of $m\angle G$ and $m\angle X$? _____

*Sketch your figure

(6) What is the perimeter of the figure? _____

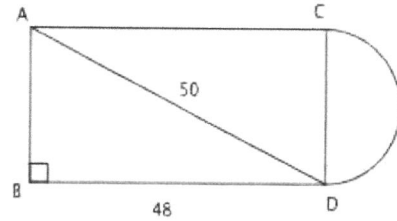

(7) Q is the center of a circle with radius $6\sqrt{3}$ cm. What is the area of the shaded region? _____

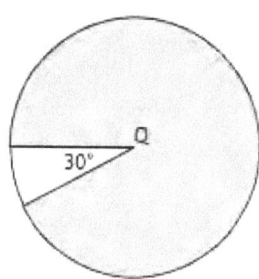

111

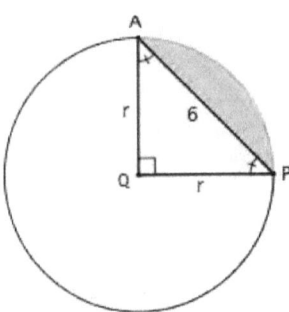

(1) Let's begin by finding the length of arc $\overset{\frown}{AP}$. We know that arc $\overset{\frown}{AP}$ is a portion of circle Q's circumference. The circumference of a circle with radius = r is given by the formula: $C_O = 2\pi r$, In our diagram, $\overline{QP} = \overline{QA} = r$. How do we find the measure of r? Well, we know that $\overline{AP} = 6$ which is the hypotenuse of right $\triangle AQP$. We also notice that $\triangle AQP$ is an isosceles right triangle (a right triangle with legs of equal length). Remember that an isosceles right triangles has side lengths in the ratio: $r : r : r\sqrt{2}$ (side:side:hypotenuse). In this case, our hypotenuse, $\overline{AP} = 6 = r\sqrt{2} \rightarrow \dfrac{6}{\sqrt{2}} = r$

. So, $C_{OQ} = 2\pi r = \dfrac{12\pi}{\sqrt{2}} = 26.66$. Now, arc $\overset{\frown}{AP}$ is opposite right angle, $\angle AQP = 90°$. Now, we recall that there

$360°$ in a circle. So, we solve for the length $\overset{\frown}{AP}$ by setting up a proportion: $\dfrac{part}{whole} = \dfrac{90°}{360°} = \dfrac{\overset{\frown}{AP}}{C_{OQ}}$. But, we just

got that $C_{OQ} = 26.66$. Solve for $\overset{\frown}{AP}$: $\dfrac{90}{360} = \dfrac{\overset{\frown}{AP}}{26.66} \rightarrow \overset{\frown}{AP} = 6.67$.

Now, we can find the area of the shaded region: We'll begin with the area of the circular sector AQP. This can be found by using the same proportion set-up we used to find $\overset{\frown}{AP}$. Remember: the area of a circle, with radius, r is given by the formula: $A_O = \pi r^2$. Therefore, $A_{OQ} = \pi \left(\dfrac{6}{\sqrt{2}} \right)^2 \rightarrow A_{OQ} = 18\pi$. Let's set up the proportion

and solve for $A_{sec\,AQP}$: $A_{sec\,AQP} : \dfrac{90}{360} = \dfrac{A_{sec\,AQP}}{18\pi} \rightarrow A_{sec\,AQP} = \dfrac{9}{2}\pi$. But the area of the shaded region is the

difference between the area of the circular sector AQP and the area of right triangle, $\triangle AQP$. The area of a

triangle is given by the formula: $A_\triangle = \frac{1}{2}(base)(height)$. So, $A_{\triangle AQP} = \dfrac{1}{2}\left(\dfrac{6}{\sqrt{2}}\right)\left(\dfrac{6}{\sqrt{2}}\right) \rightarrow A_{\triangle AQP} = 9$. We

know that $A_{sec\,AQP} = \dfrac{9}{2}\pi$, so we can finally find the area of the shaded region! :

$A_{shaded\,region} = A_{sec\,AQP} - A_{\triangle AQP} \rightarrow A_{shaded\,region} = \dfrac{9}{2}\pi - 9$.

(2) Let's label the trapezoid, $ABDC$ and draw a line from point D that is $\perp$ to $\overline{AC}$ at point H. This segment represents the shortest distance between $\overline{AC}$ and $\overline{BD}$, commonly known as the trapezoid's **altitude, "a"**. Note that segment "a" splits the trapezoid into two sections: a

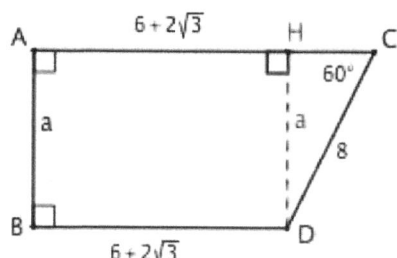

right triangle, $\triangle HDC$ and a rectangle $\square ABDH$. We can find the area of the entire figure by adding the areas of the smaller figures- $\square ABDH$ and $\triangle HDC$. We notice that we have to find the length of $\overline{HD}$ before finding our areas ($\overline{HD}$ is the height of both the rectangle and the triangle!) Because $m\angle HCD = 60°$ and $m\angle CHD = 90°$ we can conclude that $\triangle HDC$ is a $30° - 60° - 90°$ triangle. We know the ratio of the sides of a $30° - 60° - 90°$ triangle is $x : x\sqrt{3} : 2x$. Since $\overline{CD}$ is our hypotenuse (opposite angle measure $90°$) we know: $\overline{CD} = 8 = 2x$. So, we solve for x: $x = 4$. Now, $\overline{HD}$ is opposite the angle with measure $60°$. So, $\overline{HD} = x\sqrt{3} \rightarrow \overline{HD} = 4\sqrt{3}$. Note that we find the area of a rectangle by: $A_\square = (base)(height)$. For the area of a triangle, refer to question #1. Now, we can find the area of the entire figure by finding the sum of these values:

$$A_{trapABDC} = A_{\square ABDH} + A_{\triangle HDC} = \left(6 + 2\sqrt{3}\right)\left(4\sqrt{3}\right) + \frac{1}{2}\left(4\sqrt{3}\right)(4) = 79.426 \text{ or } A_{trapABDC} = 24 + 32\sqrt{3}$$

(3) We can tell that the area of the shaded region is simply the difference between the area of the square and the area of the circle. So: $A_{shadedregion} = A_\square - A_\bigcirc$. Since the radius of the circle is given at $\sqrt{7}$, we can tell that the diameter and the length of the side of the square is $2\sqrt{7}$ in. The area of a square with side s is given by the formula: $A_\square = s^2$. Therefore, $A_\square = \left(2\sqrt{7}\right)^2$ and $A_\bigcirc = 7\pi$. So $A_{shadedregion} = 28 - 7\pi \approx 6 in^2$. Now, the perimeter of the shaded region is simply the sum of the perimeter of the square and the circumference of the circle: $P_{shadedregion} = P_\bigcirc + P_\square$. Therefore, $P_{shadedregion} = 2\pi\sqrt{7} + 8\sqrt{7} = 37.79 in$

(4) We know that $\angle a$ & $\angle c$ and $\angle a$ & $\angle d$ are vertical angles, so $m\angle a = m\angle c = m\angle d$. Using substitution and the fact that three linear pairs add up to $180°$, $3m\angle a = 180° \rightarrow m\angle a = 60°$. Since $\angle a$ and the angle represented by the value of 4b are vertical angles, $m\angle a = 4b \rightarrow 60° = 4b \rightarrow b = 15°$. So, $b^2 - d = 225 - 60 = 165°$.

(5) We are given three sides of $\triangle GHX$. How do we find out the measures of its angles? We'll use the **Law of Cosines** to find $m\angle G$: $\left(HX\right)^2 = \left(GX\right)^2 + \left(HG\right)^2 - 2\left(HG\right)\left(GX\right)\cos G$ and so, $\left(17\right)^2 = \left(8\right)^2 + \left(13\right)^2 - 2\left(13\right)\left(8\right)\cos G$. Therefore, $\cos G = -.269 \rightarrow \cos^{-1}(-.269) = 105.61°$. Now we use the **Law of Sines** to find $m\angle X$. $\dfrac{HX}{\sin G} = \dfrac{HG}{\sin X} \rightarrow \dfrac{17}{.963} = \dfrac{13}{\sin X}$. So, $\sin X = .737 \rightarrow \sin^{-1}.737 = 47.43°$. Finally, we find the sum of the two angles. $m\angle G + m\angle X = 153.04°$

(6) Because we have a right triangle, $\triangle ABD$, we'll use the **Pythagorean theorem** to find the length of segment $\overline{AB}$ and the height of the rectangle: $\overline{AB} = \sqrt{50^2 - 48^2} = 14$. Since this is a rectangle, we notice that $\overline{AB} = \overline{CD} = d_{semi\bigcirc}$. (basically, $\overline{CD}$ is also the diameter of the semicircle) Therefore, $d_{semi} = 14$. Find the perimeter of the figure by finding the sum of three sides of the rectangle and the semi-circle's circumference ($C_{semi\bigcirc} = \frac{1}{2}(\pi d)$). Therefore, $P_{figure} = 2\left(48\right) + 14 + \dfrac{1}{2}\left(14\pi\right) = 131.99$.

(7) Remember that area of a circle, with radius, r, is given by the formula: $A_{\bigcirc} = \pi r^2$. Since $r = 6\sqrt{3}$, we know that $A_{circle} = \pi\left(6\sqrt{3}\right)^2$. Similar to what we did in problem one, we would like to know what portion of the circle is the shaded region. We know that the number of degrees included in this shaded region is $360° - 30° = 330°$. So, $\dfrac{part}{whole} = \dfrac{330°}{360°} = \dfrac{11}{12}$. Therefore,

$A_{shadedregion} = \dfrac{11}{12}\left(A_{circle}\right) = \dfrac{11}{12}\left(339.292\right) = 311.018\,cm^2$

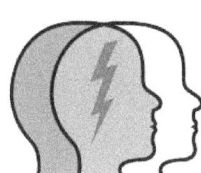
Directions: Answer the following questions by showing all your work.

Note that figures are **not** drawn to scale.

Be sure to use correct units in your answers.

handwritten:
0 17 10
180
− 96
84

(1) Find the value of x. _____ *50*

handwritten: 18F 64+70+x

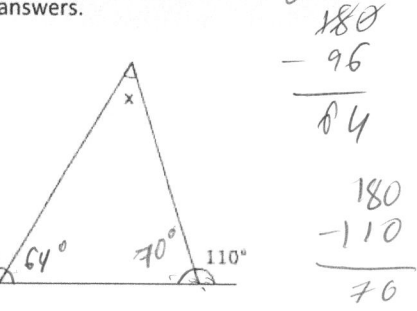

96° 64° 70° 110°

handwritten:
180
−110
70

(2) Which of the following is true? _____

a) $\overline{yz} < 8$

b) $\overline{yz} = 8$

c) $8 < \overline{yz} < 15$

d) $\overline{yz} = 15$

e) $\overline{yz} > 15$ *skiscol*

8   8°   15

60°   40°
y   z

(3) Find the value of x. _____ *60*

handwritten: 180 = 90 + 30 + x

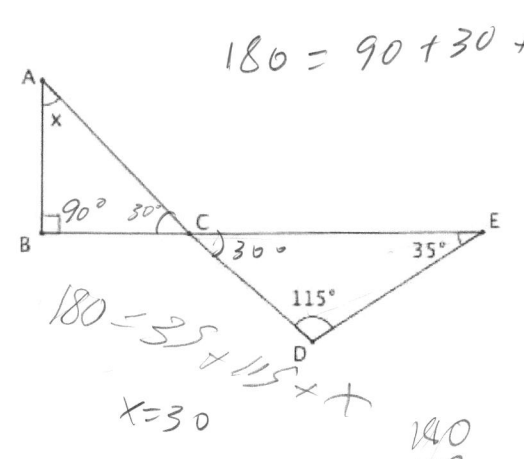

A
x
90° 30° C
B   30°   E
               35°
        115°
          D

handwritten: 180 = 35 + 115 + x

x = 30

handwritten:
180
−120
60

(4) Find the value of y. _____ *120*

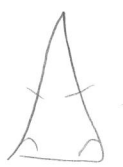

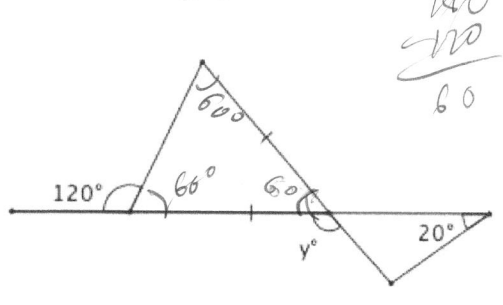

60°
120°   66°   60
                 y°   20°

handwritten:
180
−120
60

180 − 60 − 60

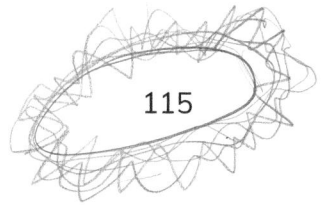

(5) Solve △BAC and △ADC (Find all ∡'s and sides).

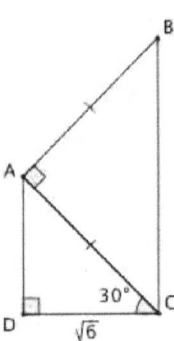

(6) Two of the lengths of the sides of a triangle are 10 and 11. Which of the following could be the length of the third side:

I. 1

II. 4

III. 22

a) None of the above b) I only c) II only d) I & II only e) All of the above

(7) What is the area of a square whose diagonal is 14? _____

(8) Solve the following triangles:

(A) A _____ $m∡b$ _____ $m∡c$ _____

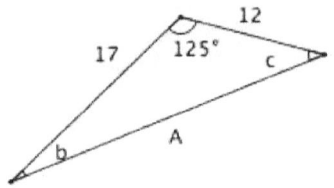

(B) $m∡a$ _____ $m∡b$ _____ $m∡c$ _____

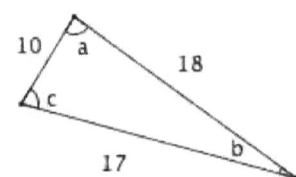

(C) A _____ B _____ $m∡b$ _____

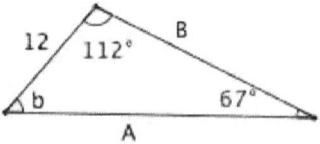

116

(9) Solve $\triangle ADC$ and $\triangle ACB$ (all $\angle$'s and sides).

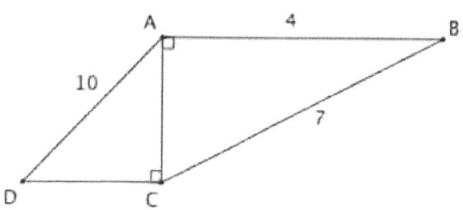

(10) Find the area of the triangle. _____

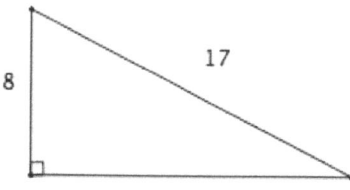

(11) Given $\overline{AB} \parallel \overline{DE}$, find the following:

$\overline{BE}$ ____ $\overline{DE}$ ____ $\overline{AB}$ ____ $m\angle x$ _____ $m\angle y$ _____ $m\angle z$

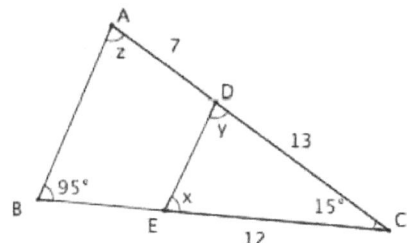

(12) In rectangle $ADCB$, $\overline{AD} = 3\sqrt{3}$. What is the perimeter _____ and the area _____ of the shaded triangle?

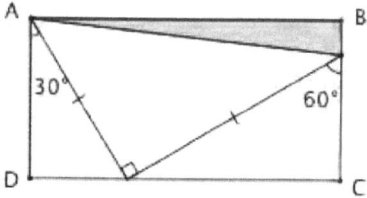

117

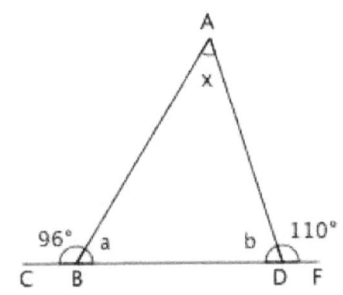

(1) Let's first find the values of a and b. Because they both form linear pairs with $\angle ABC$ and $\angle ADF$ respectively, we know that $a + m\angle ABC = 180^\circ$ and $b + m\angle ADF = 180^\circ$. Therefore, $a + 96^\circ = 180^\circ \rightarrow a = 84^\circ$ and $b + 110^\circ = 180^\circ \rightarrow b = 70^\circ$. Now, since the sum of the measures in a triangle is 180°, we can conclude that $a + b + x = 180^\circ \rightarrow 84^\circ + 70^\circ + x = 180^\circ \rightarrow x = 26^\circ$

(2) Let's begin by finding the value of $\angle x$. We know from question #1,

$$m\angle x + m\angle y + m\angle z = 180^\circ \rightarrow m\angle x + 60^\circ + 40^\circ \rightarrow m\angle x = 80^\circ.$$ Therefore, $\angle x$ is the largest angle in this triangle. Why is this so important? Well, in any triangle, a larger angle is always opposite a larger side and vice versa. In this case, yz should be greater than the other two sides. Therefore, option "e": $yz > 15$.

(3) First let's solve for the remaining angle of $\triangle ECD$, $\angle ECD$. We know that $m\angle ECD + 115^\circ + 35^\circ = 180^\circ$ (sum of the angles of a triangle). Solve: $m\angle ECD = 30^\circ$. Because $\angle ACB$ and $\angle ECD$ are vertical angles, $m\angle ECD = m\angle ACB \rightarrow m\angle ACB = 30^\circ$. But, $m\angle ACB + m\angle CAB + m\angle ABC = 180^\circ$. Therefore, $30 + x + 90 = 180 \rightarrow x = 60^\circ$.

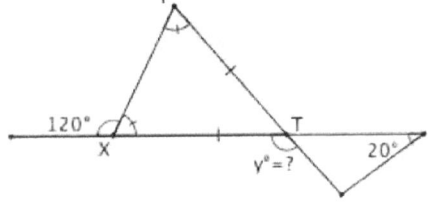

(4) Let's begin by finding the value of $m\angle PXT$. We know that the sum of linear pairs is 180°. Therefore, $120^\circ + m\angle PXT = 180^\circ \rightarrow m\angle PXT = 60^\circ$. Since $\triangle PXT$ is an isosceles $\triangle$, we know that $m\angle XPT = 60^\circ$. So, $m\angle XPT + m\angle PXT + m\angle PTX = 180^\circ$ giving us: $60^\circ + 60^\circ + m\angle PTX = 180^\circ \rightarrow m\angle PTX = 60^\circ$. But, $\angle PTX$ and y are linear pairs. So, $m\angle PTX + y = 180^\circ \rightarrow 60^\circ + y = 180^\circ \rightarrow x = 120^\circ$.

(5) Let's begin with $\triangle ADC$: The measures of its angles are 30° and 90° $m\angle DAC + 90^\circ + 30^\circ = 180^\circ \rightarrow m\angle DAC = 60^\circ$. So, we have a 30-60-90 triangle with a side ratio of $x : x\sqrt{3} : 2x$. In this case, $\sqrt{6} = x\sqrt{3} \rightarrow x = \sqrt{2} \rightarrow \sqrt{2} = \overline{AD}$ and $2x = 2\sqrt{2} = \overline{AC}$. Since $\triangle ABC$ is an isosceles right triangle, $\overline{AB} = \overline{AC} = 2\sqrt{2}$. For the angles of $\triangle ABC$, $m\angle ACB + m\angle ABC + 90^\circ = 180^\circ$. But, we know that isosceles triangles have congruent base angles, so, $2m\angle ACB + 90^\circ = 180^\circ \rightarrow m\angle ACB = 45^\circ = m\angle ABC$. The sides of an isosceles right triangle (45-45-90) are in the ratio $x : x : x\sqrt{2}$. In this case, $x = 2\sqrt{2} \rightarrow x\sqrt{2} = 4$, telling us that $\overline{BC} = 4$

(6) For this problem, we must apply the fact that the sum of the lengths of two sides of a triangle must be greater than the third side. We are given that sides of a triangle are 10 and 11. Let's call the third side "c". We know that: $11 + 10 > c \rightarrow c < 21$ and $10 + c > 11 \rightarrow c > 1$ therefore, $1 < c < 21$. So, "II. 4" is a reasonable value. Our final answer is "c": II only.

(7) The area of a square can be found in two ways: (1) $A_\square = s^2$ and (2) $A_\square = \frac{1}{2}d_1 d_2$. Because we are given the length of one of the diagonals, let's calculate the square's area using the second formula. We are given $d_1 = 14$. But, we know that the diagonals of a square are equal. So, $A_\square = \frac{1}{2}(14)(14) = 98$.

(8) (A) For this triangle, we are given the value of two sides and the included angle. Therefore, we are able to use the **Law of Cosines** to find the value of $A: A = C^2 + B^2 - 2(C)(B)\cos a$. Now, we plug in our given values and solve for A: $A^2 = 17^2 + 12^2 - 2(17)(12)\cos 125° \rightarrow A = 25.83$. Let's find c using the **Law of Sines:** $\dfrac{C}{\sin c} = \dfrac{A}{\sin a} \rightarrow \dfrac{17}{\sin c} = \dfrac{25.83}{\sin 125°}$. Solve for $\sin c$ and then use the inverse sine function to find the value of c: $\sin c = .5391 \rightarrow \sin^{-1}(.5391) = 32.63 \rightarrow c = 32.63°$. We know that: $a + b + c = 180°$ so, $125 + b + 32.63 = 180° \rightarrow b = 22.37°$.

(8) (B) For this triangle, we are given the value of three sides and no angles. Therefore, we are able to use the **Law of Cosines** to find the value of a: $A^2 = B^2 + C^2 - 2(B)(C)\cos a$. Now, we plug in our given values and solve for a: $17^2 = 18^2 + 10^2 - 2(18)(10)\cos a \rightarrow a = \cos^{-1}(.325) \rightarrow a = 67.98°$. Use the **Law of Cosines** again to find the value of c: $C^2 = A^2 + B^2 - 2(A)(B)\cos c$. Now, let's plug in our known values: $18^2 = 17^2 + 10^2 - 2(17)(10)\cos c \rightarrow c = \cos^{-1}(.191) \rightarrow c = 78.98°$. We know that: $a + b + c = 180°$ so, $67.98 + b + 78.98 = 180° \rightarrow b = 33.04°$.

(8) (C) For this triangle, we are given the value of one side and two angles. First, let's find a. We know $a + b + c = 180°$ so $112 + b + 67 = 180° \rightarrow c = 1°$. Now, we'll use the **Law of Sines** to find the values of A and B:

$$\frac{A}{\sin 112°} = \frac{12}{\sin 67°} \rightarrow A = 12.09 \text{ and } \frac{B}{\sin 1°} = \frac{12}{\sin 67°} \rightarrow B = .23.$$

(9) Because $\triangle ABC$ is a right triangle, let's use

Pythagorean theorem to find the value of $\overline{AC}$:

$BC^2 = AC^2 + AB^2 \rightarrow 7^2 = AC^2 + 4^2 \rightarrow \overline{AC} = \sqrt{33}$.

$\triangle ADC$ is also right, so let's use the Pythagorean theorem

to find the value of $\overline{DC}$:

$AD^2 = DC^2 + AC^2 \rightarrow 10^2 = DC^2 + \left(\sqrt{33}\right)^2 \rightarrow$

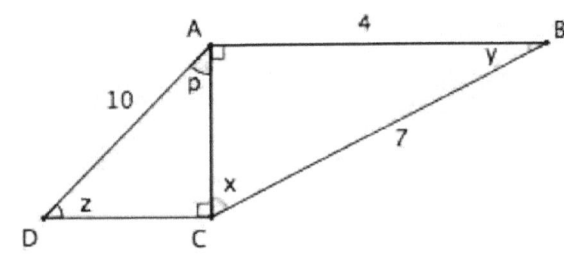

$\overline{DC} = \sqrt{67}$ We can use right triangle trigonometry and inverse trigonometric functions to find the values of the

missing angles: (*Note: remember that $\sin\angle = {}^{opp}\!/\!_{hyp}$ and $\cos\angle = {}^{adj}\!/\!_{hyp}$) So,

$\sin x = {}^4\!/\!_7 \rightarrow \sin^{-1} {}^4\!/\!_7 = x \rightarrow x = 34.85^\circ$ and $\cos y = {}^4\!/\!_7 \rightarrow \cos^{-1} {}^4\!/\!_7 = y \rightarrow y = 55.15^\circ$.

$\sin z = {}^{\sqrt{33}}\!/\!_{10} \rightarrow \sin^{-1} {}^{\sqrt{33}}\!/\!_{10} = z \rightarrow z = 35.06^\circ$ and $\cos p = {}^{\sqrt{33}}\!/\!_{10} \rightarrow \cos^{-1} {}^{\sqrt{33}}\!/\!_{10} = p \rightarrow$. $p = 54.94^\circ$

(10) Since this is a right $\triangle$, we can find the value of the third side using the Pythagorean theorem:

$17^2 = 8^2 + s^2 \rightarrow s = 15$. The area of a triangle is given by the formula: $A_\triangle = {}^1\!/\!_2 (base)(height)$. Let's

calculate by plugging in our values: $A_\triangle = {}^1\!/\!_2 (15)(8) = 60$.

(11) Since $\overline{AB} \parallel \overline{DE}$ we know that $x = 95^\circ$ because $\angle ABC$ and $\angle DEC$ are corresponding angles. We also are

sure that $x + y + 15^\circ = 180^\circ$ So, $95^\circ + y + 15^\circ = 180^\circ \rightarrow y = 70^\circ$ and $z = y \rightarrow z = 70^\circ$. because $\angle CAB$ and

$\angle CDE$ are corresponding angles. We have now proved $\triangle DCE \sim \triangle ACB$ by Angle-Angle-Angle (AAA) similarity.

Because they are similar, we know that their sides are proportionate to one another. So:

$\dfrac{\overline{AC}}{\overline{DC}} = \dfrac{\overline{BC}}{\overline{EC}} \rightarrow \dfrac{20}{13} = \dfrac{12 + \overline{BE}}{12} \rightarrow \overline{BE} = 6.46$. Now, $\overline{DE}$ can be found by using the ***Law of Sines:***

$\dfrac{\overline{DE}}{\sin 15^\circ} = \dfrac{12}{\sin 70^\circ} \rightarrow \overline{DE} = 3.31$. Now, we use the fact that $\triangle DCE \sim \triangle ACB$, so,

$\dfrac{\overline{AC}}{\overline{DC}} = \dfrac{\overline{AB}}{\overline{DE}} \rightarrow \dfrac{20}{13} = \dfrac{\overline{AB}}{3.31} \rightarrow \overline{AB} = 5.09$.

(12) We know that right triangle, $\triangle ADH$ is a 30-60-90 triangle with the
ratio of side lengths: $x : x\sqrt{3} : 2x$. So, $\overline{AD} = x\sqrt{3} = 3\sqrt{3} \to x = 3$.
Therefore, $\overline{DH} = 3$ and $\overline{AH} = 6$. But we are given $\overline{AH} = \overline{HP}$, so

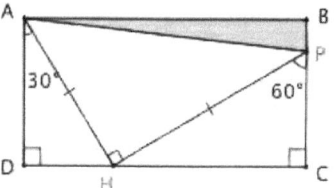

$\overline{HP} = 6$. Like $\triangle ADH$, $\triangle PHC$ is also a 30-60-90 triangle. In this case,
$\overline{HP} = 2x = 6 \to x = 3 = \overline{CP}$, and $\overline{HC} = 3\sqrt{3}$. Therefore,
$\overline{DC} = 3 + 3\sqrt{3} = \overline{AB}$ and $\overline{AD} = \overline{BC}$ (because $ADCB$ is a rectangle) so,
$3\sqrt{3} = \overline{BC} = \overline{BP} + \overline{CP} \to 3\sqrt{3} = \overline{BP} + 3 \to \overline{BP} = 3\sqrt{3} - 3$. Now, the area of $\triangle APB$ is given by the formula:
$A_{\triangle APB} = \frac{1}{2}(base)(height) = \frac{1}{2}(3 + 3\sqrt{3})(3\sqrt{3} - 3) = \frac{1}{2}(18) \to A_{\triangle APB} = 9$. Now, $\triangle AHP$ is an isosceles right
triangle, with side ratio $x : x : x\sqrt{2}$. Since $\overline{AH} = \overline{HP} = x = 6$, we know that, the hypotenuse,
$\overline{AP} = x\sqrt{2} \to \overline{AP} = 6\sqrt{2}$. So, the perimeter is the sum of the sides of the triangle: $P_{\triangle APB} = 6\sqrt{2} + 6\sqrt{3}$

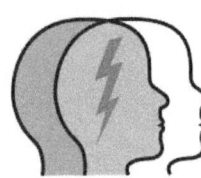

Directions: Answer the following questions by showing all your work.
Note that figures are **not** drawn to scale.
Be sure to use correct units in your answers.

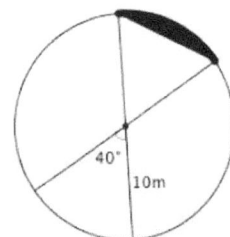

(1) What is the area of the shaded region? _____

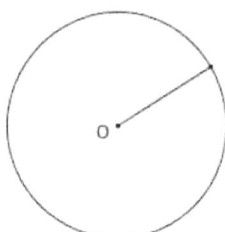

(2) If the area of circle O is 300m², find its circumference. _____

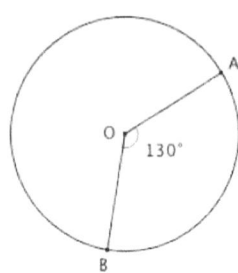

(3) Find the perimeter of sector AOB if r=8. _____

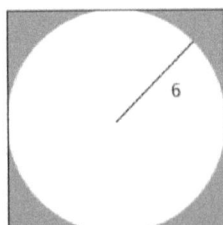

(4) Find the area of the shaded region. _____

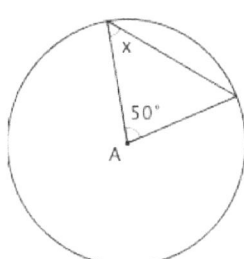

(5) Find the value of x in circle A. _____

(6) A circle is defined by: $(x - 3)^2 + (y + 7)^2 = 20$. Find: (a) center (x, y) _____ (b) radius _____

(7) Write the equation of the circle with center (-5,4) and diameter of 18. _____

(8) A is the center of a circle whose radius is 12 and B is the center of a circle whose radius is 8. If these circles are tangent to each other, what is the (a) area and (b) circumference of the circle whose diameter is $\overline{AB}$? (a) _____ (b)_____

(9) What is the area of the circle whose radius is the diagonal of a square with area 16? _____

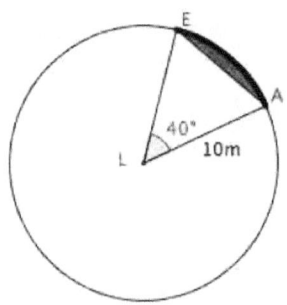

(1) The area of the shaded region in the figure is the difference between the area of the sector ELA and the area of $\triangle ELA$. First, let's find the area of sector ELA: because the sector includes only $40°$ of the circle (a circle is $360°$ around) the area of sector ELA must be a distinct portion of the area's circle. We set up a proportion: $\dfrac{part}{whole} = \dfrac{40}{360} = \dfrac{A_{sec\,ELA}}{A_{OL}}$. We use the formula for the area of a circle with radius, r to find A_{OL}: $A_O = \pi r^2 \rightarrow A_{OL} = 100\pi$. Let's plug this value into the proportion and solve for $A_{sec\,ELA}$: $\dfrac{40}{360} = \dfrac{A_{sec\,ELA}}{100\pi} \rightarrow A_{sec\,ELA} = \dfrac{100\pi}{9}$: Now, let's find the area of $\triangle ELA$:

Draw a perpendicular bisector from $\angle ELA$ to $\overline{EA}$. We'll call this line $\overline{LD}$ (not pictured). We know that since $\triangle ELA$ is an isosceles triangle, the perpendicular bisector from an angle bisects the angle and the opposite side as well. Let's use right triangle trigonometry to find the measure of $\overline{LD}$ and $\overline{DA}$: $10\cos 20° = \overline{LD}$ and $10\sin 20° = \overline{DA}$. Therefore, $\overline{EA} = 2(10\sin 20°) \rightarrow \overline{EA} = 20\sin 20°$. Now, the area of $\triangle ELA$ is given by:

$A_{\triangle ELA} = \frac{1}{2}(base)(height) = \frac{1}{2}(\overline{EA})(\overline{LD}) = \frac{1}{2}(20\sin 20°)(10\cos 20°) = 32.14$

$A_{shaded\,region} = A_{sec\,ELA} - A_{\triangle ELA} = \dfrac{100\pi}{9} - 32.14 = 2.77 m^2$

(2) The area of a circle is given by the formula $A_O = \pi r^2$. In this case, $A_o = 300 = \pi r^2$. Solve for r:

$r = \sqrt{\dfrac{300}{\pi}}$ The circumference of a circle is given by the formula: $C_O = 2\pi r$. Substitute for "r" and calculate:

$C_O = 2\pi\sqrt{\dfrac{300}{\pi}} = 61.4 m$ or $C_O = 20\pi\sqrt{3}$

(3) Similar to question #1, we want to find what **portion** of the circle is sector AOB. But, we'll approach this question in a difference way. Since sector AOB includes an angle of $130°$, $\overparen{AB}$ is $\dfrac{130°}{360°} = \dfrac{13}{36}$ th of the circle's entire circumference. Since $\overline{OA} = \overline{OB} = r = 8$ and $C_{circle\,O} = 2\pi r$ we have $C_O = 2\pi r = 16\pi$ So, $\overparen{AB} = \dfrac{13}{36}(C_O) = \dfrac{52}{9}\pi$. Now, the perimeter of sector AOB equals the sum of the measures of $\overparen{AB}$, $\overline{OA}$, and $\overline{OB}$. Therefore, $P_{sec\,AOB} = \overline{OA} + \overline{OB} + m\overparen{AB}$. Let's plug in our values: $P_{sec\,AOB} = 8 + 8 + \dfrac{52}{9}\pi = 34.15$.

(4) The area of the shaded region in the diagram is equal to the difference between the area of the square and the area of the circle. We are given $r = 6$.

Since $d = 2r$, $d = 2(6) = 12 \rightarrow \overline{AB} = 12$. We also notice that $\overline{AB} = s$, a side of the square. The area of a square is given by the formula $A_\square = s^2$ So, $A_\square = 12^2 = 144$. Now, the area of a circle is given by the formula: $A_O = \pi r^2$

So, $A_O = 6^2 \pi = 36\pi$. Finally,

$A_{shaded region} = A_\square - A_O = 144 - 36\pi \rightarrow A_{shaded region} = 30.9$.

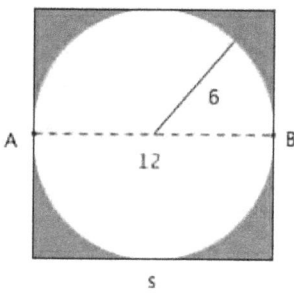

(5) First of all, we know that $\triangle ABC$ is an isosceles triangle because

$\overline{AB} = \overline{BC} = r$, Therefore, $m\angle BCA = m\angle BAC = x$ Because the sum of the angles of a triangle is $180°$, $m\angle BCA + x + 50° = 180°$. So, using substitution:

$2x + 50° = 180° \rightarrow 2x = 130° \rightarrow x = 65°$

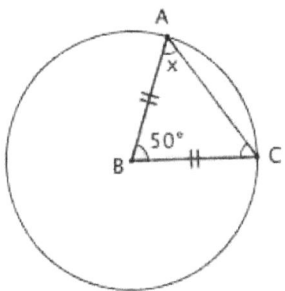

(6) The equation of a circle is in the form: $(x - h)^2 + (y - k)^2 = r^2$ where (h,k) represents the center of the circle, and $r = radius$. Therefore, if a circle is defined by $(x - 3)^2 + (y + 7)^2 = 20$, $(h,k) = (3,-7)$ and

$r^2 = 20 \rightarrow r = 2\sqrt{5}$.

(7) Similar to question #6, the equation of a circle is: $(x - h)^2 + (y - k)^2 = r^2$. If the center of this circle is $(-5,4)$, then $(h,k) = (-5,4)$. Also, the diameter is twice the radius: $d = 2r$. So,

$d = 2r \rightarrow 18 = 2r \rightarrow r = 9$ Therefore, the equation of this circle is: $(x + 5)^2 + (y - 4)^2 = 81$.

(8) First draw circle A and circle B so that they are tangent to each other and that $r_A = 8$ and $r_B = 12$. Now, draw $\overline{AB}$. Notice that

$\overline{AB} = r_A + r_B \rightarrow \overline{AB} = 20$. Therefore, if there was a circle with $diameter = 20$, we know that the diameter is twice the circle's radius. So, $d = 2r \rightarrow 20 = 2r \rightarrow r = 10$. So, the area of this circle is: $A_{O_{r10}} = \pi r^2 \rightarrow A_{O_{r10}} = 100\pi$. Since circumference, $C = 2\pi r$. $C_{O_{r10}} = 20\pi$.

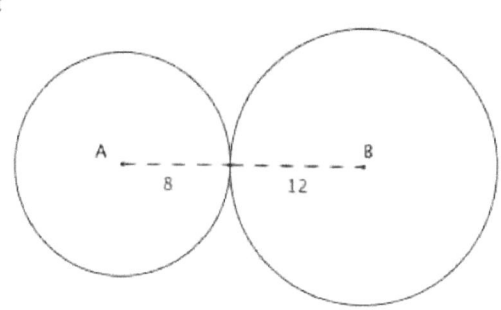

(9) Let's first start with the square: the area of a square is given by the formula: $A_{\square} = s^2$. Now, let's use the given information to solve for s, the side of the square: $16 = s^2 \rightarrow s = 4$. If you draw a diagonal of a square, you form a 45-45-90 triangle, also known as a isosceles right triangle. The sides of this kind of triangle is $x : x : x\sqrt{2}$. In this case, $d = 4\sqrt{2}$. Back to the question: we assume $r = d = 4\sqrt{2}$. The area of this circle is given by the formula: $A_O = \pi r^2$, so using our radius, $A_O = \pi(4\sqrt{2})^2 \rightarrow A_O = 32\pi$.

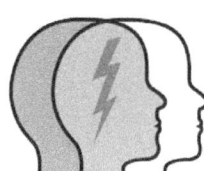

Directions: Answer the following questions by showing all your work.
Note that figures are ***not*** drawn to scale.
Be sure to use correct units in your answers.

(1) How many diagonals are there in a/an (a) octagon? _____ (b) dodecagon? _____ (c) 92-gon? _____

(2) Find the area of trapezoid $BADC$. _____

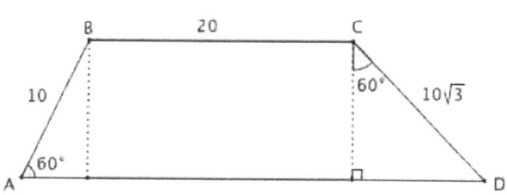

(3) Find the area of rhombus $YXWZ$. _____

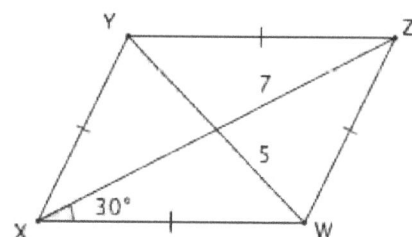

(4) Find the area of parallelogram $NMPO$. _____

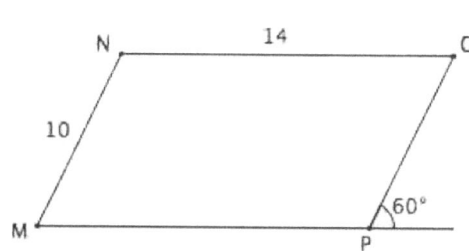

(5) What is the measure of an interior angle of a regular nonagon? _____ A 57-gon? _____

(6) Find the area of the shaded region within this regular

polygon. _____

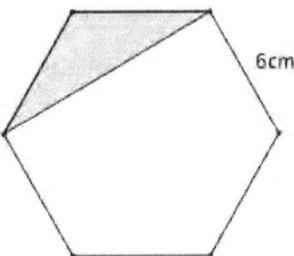

6cm

(7) Find the volume of the cylinder with radius of 4 and a diagonal length of 17. _____

(8) Find the volume of the cube. _____

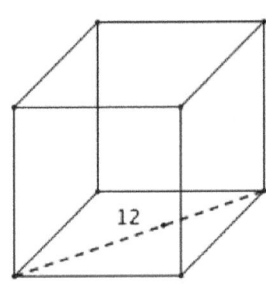

12

(9) $\overline{GD}$ is a diagonal that runs through the center of the

cube. What's its length? _____

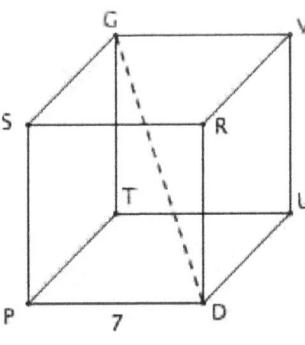

(10) Find the area of the square. _____

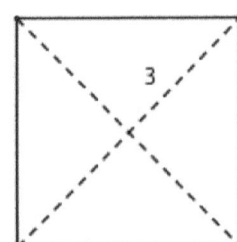

3

(1) To find the amount of diagonals of a polygon with n number of sides (***note**-this can also be the number of angles), use the formula: $D = \left(\dfrac{n}{2}\right)(n-3)$ where D represents the number of diagonals. Now, (a) An octagon has 8 sides so $D_{oct} = \left(\dfrac{8}{2}\right)(8-3) = 20$ (b) A dodecagon has 12 sides, so

$D_{dodec} = \left(\dfrac{12}{2}\right)(12-3) = 54$ (c) There are 92 sides in a 92-gon so $D_{92} = \left(\dfrac{92}{2}\right)(92-3) = 4094$.

(2) To find the area of trapezoid $BADC$, we'll split the diagram into three parts: $\triangle BAN$, $\square BNMC$ and $\triangle CMD$. First, let's find the height of $\triangle CMD$, segment $-\overline{CM}$. Since $\angle MCD = 60°$ and $\angle CMD = 90°$ we know that we have a 30-

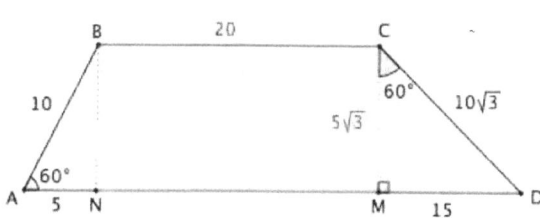

60-90 triangle with side lengths in the ratio of: $x : x\sqrt{3} : 2x$. In this case, $10\sqrt{3} = 2x \rightarrow x = 5\sqrt{3} = \overline{CM}$. And, we know that $MD = x\sqrt{3} = (5\sqrt{3})(\sqrt{3}) = 15$. The area of a triangle is given by the formula:

$A_\triangle = \frac{1}{2}(base)(height)$, so $A_{\triangle CMD} = \frac{1}{2}(15)(5\sqrt{3}) \rightarrow A_{\triangle CMD} = \dfrac{75\sqrt{3}}{2}$. Note that the height of $\triangle CMD$

is also the height of $\square BNMC$. Now, the area of a rectangle is given by the formula: $A_\square = base \times height$, so

$A_{\square BNMC} = (20)(5\sqrt{3}) \rightarrow A_{\square BNMC} = 100\sqrt{3}$. Now, let's find the base length, $\overline{AN}$ of $\triangle ANB$. Again, we can

tell that $\triangle ANB$ is a 30-60-90 triangle with $\overline{AB} = 10 = 2x$, therefore, $\overline{AN} = x = 5$. We also know that the

triangle's height is $\overline{NB} = 5\sqrt{3}$. So, $A_{\triangle BAN} = \frac{1}{2}(5)(5\sqrt{3}) \rightarrow A_{\triangle BAN} = \dfrac{25\sqrt{3}}{2}$. Therefore,

$A_{trapBADC} = A_{\triangle CMD} + A_{\square BNMC} + A_{\triangle BAN} \rightarrow A_{trapBADC} = 150\sqrt{3}$

(3) Let's begin by finding little bit more information about rhombus $YXWZ$. Since the diagonals of a rhombus bisect each other we know that $d_1 = \overline{ZX} = 7(2) \rightarrow d_1 = 14$ and $d_2 = \overline{YW} = 5(2) \rightarrow d_2 = 10$. Now, we know that the area of a rhombus is given by the formula: $A_\square = \frac{1}{2}d_1d_2$. Therefore,

$A_{rbYXWZ} = \frac{1}{2}(14)(10) \rightarrow A_{rbYXWZ} = 70$.

(4) The area of a parallelogram is given by the formula, $A_{\square} = (base)(height)$. In the case of a parallelogram like $\square NMPO$, the height is the polygon's altitude: the shortest length between the top and bottom segments of the parallelogram, in this case, $\overline{MP}$ and $\overline{NO}$. Now, we draw

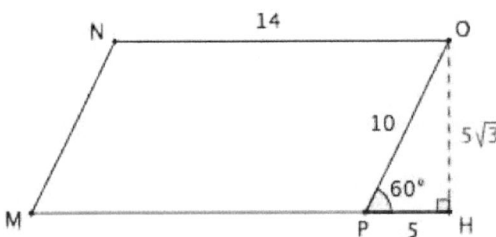

altitude $\overline{OH}$, creating right triangle $\triangle OPH$. (*Note- you can also draw the altitude INSIDE the parallelogram. Considering the information given, it is much simpler to draw it on the outside). We see that $\triangle OPH$ is a 30-60-90 triangle with a the ratio of sides: $x : x\sqrt{3} : 2x$. In this case, $\overline{PO} = 2x = 10 \rightarrow x = 5$. Therefore, altitude = $\overline{OH} = 5\sqrt{3}$. $A_{\square NMPO} = (14)(5\sqrt{3}) \rightarrow A_{\square NMPO} = 70\sqrt{3}$

(5) The measure of an interior angle of a regular polygon with 'n' number of sides is given by the formula: $m\angle \text{int} = \dfrac{180(n-2)}{n}$. We know that a nonagon has 9 sides. Therefore, $m\angle \text{int}_{nonagon} = \dfrac{180(9-2)}{9} = 140°$. And, we know that a 57-gon has 57 sides. Therefore, $m\angle \text{int}_{57-gon} = \dfrac{180(57-2)}{57} = 173.68°$

(6) We want to find the area of the shaded region, $\triangle ABC$. Since it is included in a regular hexagon, we know that $\overline{BA} = \overline{AC} = 6$. We also know (from question #5) that the measure of an interior angle of a regular hexagon is: $m\angle \text{int}_{hexagon} = \dfrac{180(6-2)}{6} = 120°$. Therefore,

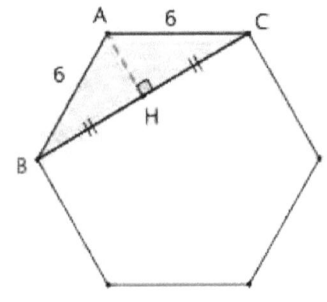

$m\angle BAC = 120°$. Now, we find the area of this isosceles triangle by drawing the perpendicular bisector from $\angle BAC$ to $\overline{BC}$. Because $\triangle ABC$ is isosceles, we know that the perpendicular bisector, $\overline{AH}$ also bisects $\angle BAC$. Therefore, $m\angle BAH = 60° = m\angle CAH$ .. But, we also know that $\overline{AH} = \overline{AH}$ and $m\angle AHB = 90° = m\angle AHC$, which tells us that $\triangle CAH \cong \triangle BAH$ (and 30-60-90 triangles). To find $\overline{AH}$, which is our height of the triangle, we use the side ratio of a 30-60-90 triangle: $x : x\sqrt{3} : 2x$. So, in $\triangle CAH$, $\overline{AC} = 6 = 2x \rightarrow x = 3 = \overline{AH}$. And, $\overline{BH} = \overline{HC} = x\sqrt{3} = 3\sqrt{3}$. So, $\overline{BC} = 2\overline{BH} \rightarrow \overline{BC} = 6\sqrt{3}$, which is the measure of our base. We know that the area of a triangle is given by the formula: $A_{\triangle} = \frac{1}{2}(base \times height)$, so $A_{\triangle ABC} = \frac{1}{2}(6\sqrt{3})(3) \rightarrow A_{shadedregion} = A_{\triangle ABC} = 9\sqrt{3}$.

(7) The volume of a cylinder is given by the formula: $V_{cylinder} = \pi r^2 h$, where 'r' represents the radius of the cylinder's base. In this case, $r = 4$. Now, we must find this cylinder's height! First, let's draw the cylinder. In the diagram to the right, we see that $r = \overline{NP} = 4$ Draw $\overline{AP}$. Therefore, the diameter, $\overline{AP} = 2r = 8$. Now, are given that the diagonal of the cylinder, $\overline{PL} = 17$. So, we have right $\triangle PLA$, with side 8 and hypotenuse, 17-meaning our last side is 15! (right triangle proportions 8-15-17). Therefore, $\overline{LA}$, the height of $\triangle PLA$ is 15. Now,

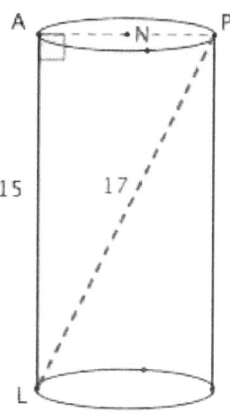

$$V_{cylinder} = \pi(4)^2(15) \rightarrow V_{cylinder} = \pi(4)^2(15) \rightarrow 240\pi$$

(8) The volume of a cube is given by the formula: $V_{cube} = s^3$, where 's' represents a side of the cube. Now, we are given that the diagonal of the base, $\overline{XY} = 12$. Since all the faces of a cube are squares, we know that that the base is a square (*see figure*) and $\overline{QY} = \overline{QX} = s$. Therefore, we have an isosceles right $\triangle XYQ$ with side ratio: $s : s : s\sqrt{2}$. In this case,

base (cube)

$$XY = 12 = s\sqrt{2} \rightarrow s = \frac{12}{\sqrt{2}} \cdot V_{cube} = \left(\frac{12}{\sqrt{2}}\right)^3 \rightarrow 432\sqrt{2}$$

(9) We begin to find the length of diagonal $\overline{GD}$ by creating right $\triangle GTD$ consisting of diagonal $\overline{GD}$, side $\overline{GT}$ and base diagonal, $\overline{DT}$. Since we know $\overline{GT} = s_{cube} = 7$, we must find the length of $\overline{DT}$. Similar to question #8, we can the diagonal of the base of a cube (square) creates an isosceles right triangle with the sides of the cube. In this case, we have right $\triangle TDU$. The sides in these triangles are in a $s : s : s\sqrt{2}$ proportion. Here, $\overline{DU} = \overline{TU} = s = 7$. Therefore, $\overline{DT} = 7\sqrt{2}$. So, we'll use Pythagorean theorem to find $\overline{GD}$:

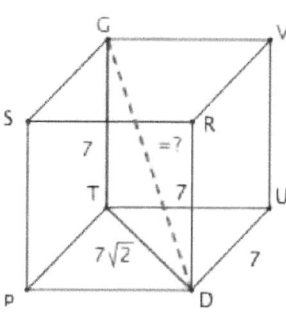

$$GD^2 = DT^2 + GT^2 \rightarrow GD = \sqrt{98 + 49} \rightarrow \overline{GD} = 12.12 \text{ or } \overline{GD} = 7\sqrt{3}$$

(10) Because the diagonals of a square bisect each other we know that $d_1 = 2(3) = 6$. Now, the area of a square can be found in two ways: (1) $A_{\square} = s^2$ and $A_{\square} = \frac{1}{2}d_1 d_2$ (2) Because we are given the length of one of the diagonals, let's calculate the square's area using the second formula. We are given $d_1 = 6$ But, we know that the diagonals of a square are equal. So, $A_{\square} = \frac{1}{2}(6)(6) \rightarrow A_{\square} = 18$.

131

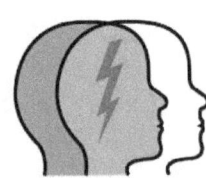

COORDINATE GEOMETRY

Directions: Answer the following questions by showing all your work.

Note that figures are *not* drawn to scale.

Be sure to use correct units in your answers.

(1) What is the distance between $A(-3,8)$ and $B(2,-4)$? _____ What are the coordinates of midpoint

M between A and B? _____ What is the slope of the line passing through all three points? _____

(2) Find the perimeter of the figure. _____

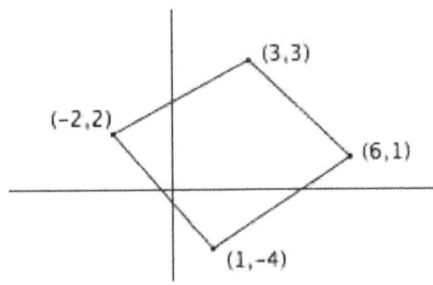

(3) What is the area of the figure bounded by the lines: $y = 3$, $x = 7$ and $y = \frac{2}{3}x + 3$? _____

(4) $\overline{JQ}$ represents this circle's diameter. What is the circle's

equation? _____

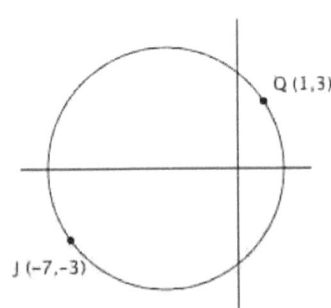

132

(5) What is the slope of the line that is perpendicular to $\overleftrightarrow{AB}$? _____

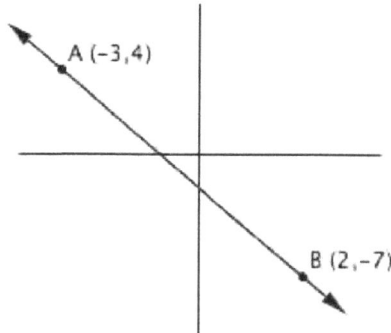

A (−3,4)

B (2,−7)

(6) Segment $\overline{JR}$ includes points $J(-1,5)$, $R(-3,k)$, and midpoint $M(x,7)$. What is the value of $k - x$? _____

(7) What is the x-intercept of the line with equation $y = \frac{4}{7}x - 5$? _____

(8) At how many points do these lines intersect? _____

$5x - 9y = 13$

$35x - 63y = 91$

(1) To find the distance between $A(-3,8)$ and $B(2,-4)$, use the distance formula:

$d = \sqrt{(x_2 - x_1)^2 + (y_2 - y_1)^2}$. In this case, $(x_1,y_1) = (-3,8)$ and. $(x_2,y_2) = (2,-4)$. Now, distance,

$d = \sqrt{(2-(-3))^2 + (-4-8)^2} = 13$. Let's find the midpoint, $M(x_M,y_M)$: x-coordinate:

$x_M = \dfrac{x_2 + x_1}{2} = \dfrac{-3+2}{2} = -\dfrac{1}{2}$ and the y-coordinate: $y_M = \dfrac{y_2 + y_1}{2} = \dfrac{8-4}{2} = 2$. Therefore,

$M(x_M,y_M) = (-\dfrac{1}{2},2)$. The slope of the line that passes through two points is given by the formula (m=slope)

$m = \dfrac{y_2 - y_1}{x_2 - x_1}$. Therefore, the slope between A and B is: $m = \dfrac{-4-8}{2-(-3)} \rightarrow m = \dfrac{-12}{5}$.

(2) To find the perimeter of the figure we must find all four distances between each of the points using the distance formula discussed in question #1. Then, we'll calculate the perimeter by adding these distances:

$d_{AB} = \sqrt{(3-2)^2 + (3-(-2))^2} = \sqrt{26}$, $d_{BC} = \sqrt{(1-(-2))^2 + (-4-2)^2} = \sqrt{45} = 3\sqrt{5}$,

$d_{CD} = \sqrt{(6-1)^2 + (1-(-4))^2} = \sqrt{50} = 5\sqrt{2}$ and $d_{DA} = \sqrt{(6-3)^2 + (1-3)^2} = \sqrt{13}$.

$P = d_{AB} + d_{BC} + d_{CD} + d_{DA} = \sqrt{26} + \sqrt{13} + 3\sqrt{5} + 5\sqrt{2} = 22.48$

So, $P_{figure} = \sqrt{26} + \sqrt{13} + 3\sqrt{5} + 5\sqrt{2}$ or $P_{figure} = 22.48$

(3) To find the points that bound this figure, we need to find its three intersection points let's call them A, B, and C. Point A is the intersection between $y = 3$ and $y = \frac{2}{3}x + 3$. Since these lines intersect in one point, A, let's set these two equations equal to each other and solve for x_A $3 = \frac{2}{3}x_A + 3 \rightarrow x_A = 0$: Since $y = 3$, $y_A = 3$. Therefore, $A(0,3)$. Now,

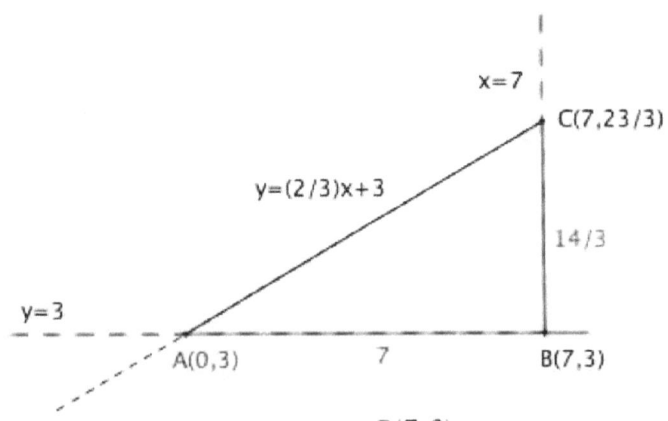

point B is the intersection between $y = 3 \rightarrow y_B = 3$ and $x = 7 \rightarrow x_B = 7$. Therefore, $B(7,3)$ Now, point C is the intersections between $y = \frac{2}{3}x + 3$ and $x = 7 \rightarrow x_C = 7$. We'll substitute $x_C = 7$ and solve for y_C:

$y_C = \frac{2}{3}(7) + 3 \rightarrow y_c = \frac{23}{3}$ And so, $C(7,\frac{23}{3})$. As you can see we have formed a right triangle.: $\overline{AB} = 7$

and $\overline{BC} = \frac{14}{3}$. And the area of $\triangle CAB$ is given by the formula:

$A_{\triangle CAB} = \frac{1}{2}(base)(height) = \frac{1}{2}(7)(\frac{14}{3}) \rightarrow A_{\triangle CAB} = \frac{49}{3}$.

(4) Let's begin by finding the measure of $\overline{JQ}$ (the circle's diameter), using the distance formula:

$d = \sqrt{(x_2 - x_1)^2 + (y_2 - y_1)^2}$. Therefore, $d_{\overline{JQ}} = \sqrt{(3-(-3))^2 + (1-(-7))^2} = \sqrt{100} = 10$. We also already

know that the radius of this circle is half of this, so $r = 5$. Now, the center of the circle, C, is the midpoint

between J and Q. So, we use what we know about the midpoint formula:

$x_M = \dfrac{x_2 + x_1}{2} = \dfrac{1 + (-7)}{2} \to x_M = -3$ and $y_M = \dfrac{y_2 + y_1}{2} = \dfrac{-3 + 3}{2} \to y_M = 0$. Therefore, $C(x_M, y_M) = (-3, 0)$

The typical equation of a circle is: $(x - h)^2 + (y - k)^2 = r^2$ where (h, k) represents the circle's center, and

r=radius. Now, we can substitute the appropriate values to find the equation of this circle: $(x + 3)^2 + y^2 = 25$.

(5) We begin by finding the slope of this line. From question #1, we know that the formula for the slope of a line

(m=slope) that goes through two points is given by: $m = \dfrac{y_2 - y_1}{x_2 - x_1}$. If $A(-3, 4) = (x_1, y_1)$ and

$B(2, -7) = (x_2, y_2)$, we can conclude: $m = \dfrac{-7 - 4}{2 - (-3)} \to m = \dfrac{-11}{5}$. Now, the slope of the perpendicular line is

the "negative reciprocal" of the original line. (to take the "negative reciprocal" you must change the sign and

switch the numerator and denominator of slope of the original line) . In this case, $m_\perp = \dfrac{5}{11}$.

(6) We are given the midpoint $M(x, 7)$ between points $J(-1, 5)$ and $R(-3, k)$. We want to use our formula

for midpoint as explained in question #1 to find the values of x and k. We have $M(x, 7) = (x_M, y_M)$. So, we

can solve for **x**: $x_M = \dfrac{(-1) + (-3)}{2} \to x_M = -2 \to x = -2$ and for **k**: $y_M = \dfrac{5 + k}{2} \to k = 9$ Therefore,

$k - x = 9 - (-2) = 11$.

(7) The x-intercept of any line is the point where the **_line passes through the x-axis_**. At this point, $y = 0$ and

$x = n$, where n is any real number ($n \in \mathbb{R}$). So, we find the x-intercept of the line given by setting y=0 and

solving for x.: $y = 0 = \frac{4}{7}x - 5 \to x = \frac{35}{4}$. Therefore, our x-intercept is: $\left(\frac{35}{4}, 0\right)$.

(8) Don't be tricked by this! These two lines intersect in *infinitely many points*. Why? Because they are the

same line! Look: $7(5x) - 7(9y) = 7(13) \to 35x - 63y = 91$, which is our second equation. As you can see,

the second equation is simply a multiple (of 7) of the first equation.

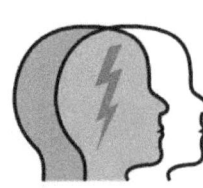

Directions: Answer the following questions by showing all your work. When possible, avoid using your calculator.

Simplify (if possible):

1) $2a^3 - 2a^2$ ← diff powers
can't simplify

2) $\dfrac{(x^2y^3)(x^{-3})}{x^{-6}} + 4y^3$

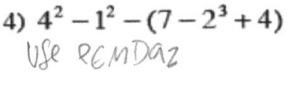
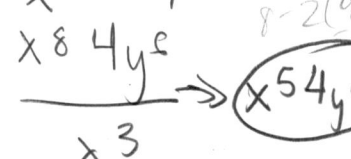

$\dfrac{x^2y^3x^6}{x^3} + \dfrac{4y^3}{1}$

3) $(4z^4 - 7)(z + 2)$

$\dfrac{x \cdot x}{x \cdot x} \cdot x = \dfrac{1}{x}$

$4z^5 + 8z^4 - 7z - 14$ $\dfrac{x^8 y^3}{x^3} + \dfrac{4y^3}{1}$

4) $4^2 - 1^2 - (7 - 2^3 + 4)$
use PEMDAS

$\dfrac{x^8 \, 4y^6}{x^3} \Rightarrow \boxed{x^5 4 y^6}$

5) $(x^n)(6x^{n+3})$

6) $5y^c + y^b$

7) $(r^t)^b + (xy)^2$

8) $8 - 2(3^2 + 6 \cdot 7) + \dfrac{1}{8}$

$8 - 2(9 + 42) + \dfrac{1}{8}$

Solve for x (find all possible values):

9) $12 - 3x + 6x = -33$

10) $bx - w + zt = z + 7bx$

11) $(2x + 5) - 5 < 7 - 5$

12) $-2 - 9|2x - 3| > -209$

$+2 \qquad\qquad +2$

$\dfrac{-9|2x - 3|}{-9} > \dfrac{-207}{-9}$

to find all sol

$|2x - 3| < 23$

13) $7 - 12x \leq -113$
$\quad -7 \qquad -7$

14) $x\sqrt{a^2 + b^2} = -7x + 9bx$

15) $|2x| - 5 < 5$
$\quad +5 \quad +5$
$|2x| < 10$

16) $-6 - 4|3 - x| \geq 100$

 glup glup

$2x - 3 < 23 \qquad 2x - 3 > 23$
$\quad +3 \quad +3$
$\dfrac{2x}{2} < \dfrac{23}{2}$ $\boxed{x > -10}$

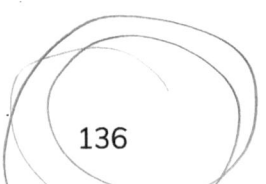

$\boxed{x < 13} \; -10 < x < 13$

136

1) Don't be fooled by this! $2a^3 - 2a^2$ *cannot be simplified* anymore because it does not include **like terms**!

2) We begin by simplifying the first term of the expression: $\dfrac{(x^2 y^3)(x^{-3})}{x^{-6}} = \dfrac{x^{-1} y^3}{x^{-6}} = x^{-1-(-6)} y^3 = x^5 y^3$. Note that we used the law of exponents: $\dfrac{x^a}{x^b} = x^{a-b}$. Now, we have: $x^5 y^3 + 4y^3$.

3) Let's begin by multiplying the expressions: $(4z^4 - 7)(z + 2) = 4z^5 + 8z^4 - 7z - 14$. We cannot simplify anymore because there are no like terms in the final expression.

4) Simplify by working on one term at a time: $4^2 - 1^2 - (7 - 2^3 + 4) = 16 - 1 - 7 + 8 - 4 = 12$

5) To simplify this expression we use the law of exponents that: $(x^a)(x^b) = x^{a+b}$. Therefore,

$(x^n)(6x^{n+3}) = 6x^{2n+3}$

6) Don't be fooled! We cannot simplify this expression by adding- there are no like terms!

7) We use our law of exponents to simplify: $(r^t)^b + (xy)^2 = r^{tb} + x^2 y^2$

8) Simplify by working on one term at a time:

$8 - 2(3^2 + 6 \cdot 7) + \dfrac{1}{8} = 8 - 2(9 + 42) + \dfrac{1}{8} = 8 - 102 + \dfrac{1}{8} \rightarrow -93.875$

9) We solve for x by isolating the variable and dividing: $12 - 3x + 6x = -33 \rightarrow 3x = -45 \rightarrow x = -15$

10) $bx - w + zt = z + 7bx \rightarrow 6bx = -w + zt - z \rightarrow x = \dfrac{-w + zt - z}{6b}$.

11) We solve for x as we would if the "<" were a "=" sign: $(2x + 5) - 5 < 7 - 5 \rightarrow x < 1$.

12) Our first step in solving an inequality with an absolute value is to isolate the absolute value on one side of the equation like so: $-2 - 9|2x - 3| > -209 \rightarrow -9|2x - 3| > -207 \rightarrow -|2x - 3| > -23$. Now, we want to divide by -1: which will, in fact, switch "<" to ">": $-|2x - 3| > -23 \rightarrow |2x - 3| < 23$. This shows us that: $2x - 3 < 23$ and $2x - 3 > -23$. Therefore, $-10 < x < 13$.

13) $7 - 12x \le -113 \rightarrow -12x \le -120 \rightarrow x \ge 10$

14) We begin to solve for x by bringing all "x-terms" to one side of the equation- and factoring out x:

$$x\sqrt{a^2 + b^2} = -7x + 9bx \rightarrow x\sqrt{a^2 + b^2} + 7x - 9bx = 0 \rightarrow x(\sqrt{a^2 + b^2} + 7 - 9b) = 0.$$ Now, divide both sides

by $\sqrt{a^2 + b^2} + 7 - 9b$. We conclude that $x = 0$.

15) We are going to use an approach, similar to that of question #12: Let's begin by isolating the absolute value:

$|2x| - 5 < 5 \rightarrow |2x| < 10$. So, the value of $2x$ is between -10 and 10: $-10 < 2x < 10 \rightarrow -5 < x < 5$

16) We begin by isolating the absolute value: $-6 - 4|3 - x| \ge 100 \rightarrow -4|3 - x| \ge 106 \rightarrow |3 - x| \le -26.5$;

(*Reminder: when you divide an inequality through by "-1", you must change the sign of the inequality). Now, we

know that $3 - x \le -26.5 \rightarrow -x \le -29.5 \rightarrow x \ge 29.5$ **and** $3 - x \ge 26.5 \rightarrow -x \ge 23.5 \rightarrow x \le -23.5$. Therefore,

the answer is *null-set (no solution)* because no value can be greater than 29.5 **and** less than -23.5. (**Note**: you can

also tell that there is no solution earlier in the solving process. You know that $|3 - x|$ is always > 0, therefore

it is impossible that $|3 - x| \le -26.5$, and so there is no solution for x.)

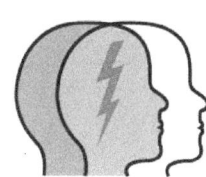

Directions: Answer the following questions by showing all your work. When possible, avoid using your calculator.

Expand:

1) $(6+\sqrt{3})\left(\dfrac{1}{2}-4\sqrt{2x}\right)$

2) $(11x+i)^2$

3) $(x^2+\sqrt{-3})(x^2-\sqrt{-2})$

Simplify:

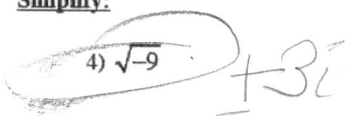

4) $\sqrt{-9}$ $+3i$

5) $\sqrt{-x}$

6) $3i^{87}$

Solve for x (find all possible values):

7) $x^2+3x-7=3$

8) $9x^2-25=0$

9) $5x^2=7x-6$

10) $2x^2-9x=-\dfrac{1}{2}-6x$

11) $-x^2=-2x-3$

12) $6x+13x^2-12=3x-15+12x^2$

139

1) We expand this equation by performing multiplication. You can do this in a variety of correct ways, but we'll use method we call FOIL- multiply the **first** two terms, the **outer** two terms, the **inner** two terms, and then the **last** two terms. Finally, combine 'like terms' to simplify:

$$(6 + \sqrt{3})\left(\frac{1}{2} - 4\sqrt{2x}\right) = 6\left(\frac{1}{2}\right) - 24(\sqrt{2x}) + \sqrt{3}\left(\frac{1}{2}\right) - 4\sqrt{6x} = 3 - 24(\sqrt{2x}) - 4\sqrt{6x} + \frac{\sqrt{3}}{2}$$

2) We know that: $(11x + i)^2 = (11x + i)(11x + i)$. Now, how do we expand? We can use the FOIL method described above: $(11x + i)(11x + i) = 11x(11x) + 11x(i) + 11x(i) + i(i) = 121x^2 + 121x + 22xi + i^2$. We can simplify this further because we know that $i^2 = -1$. Therefore, $121x^2 + 22xi + i^2 \rightarrow 121x^2 + 22xi - 1$.

3) Similar to questions #1/#2, we're multiplying binomials. We'll expand by FOIL-ing and simplifying:

$(x^2 + \sqrt{-3})(x^2 - \sqrt{-2}) = x^2(x^2) - x^2\sqrt{-2} + x^2\sqrt{-3} - \sqrt{6}$. But, $\sqrt{-3} = i\sqrt{3}$ and $\sqrt{-2} = i\sqrt{2}$ (see question #4). Therefore, $x^2(x^2) - x^2\sqrt{-2} + x^2\sqrt{-3} - \sqrt{6} \rightarrow x^4 - x^2 i\sqrt{2} + x^2 i\sqrt{3} - \sqrt{6}$

4) We have: $\sqrt{-9} = \sqrt{9 * -1} = \sqrt{9 * i^2}$ because $i^2 = -1$. Therefore, $\sqrt{9 * i^2} = \sqrt{9} * \sqrt{i^2} = \pm 3i$.

5) We have: $\sqrt{-x} = \sqrt{x \cdot -1} = \sqrt{x \cdot i^2}$ because $i^2 = -1$, Therefore, $\sqrt{x \cdot i^2} = \sqrt{x}\sqrt{i^2} = i\sqrt{x}$.

6) Let's begin simplifying this term by stating the facts: Since $i^2 = -1$ we know that:

$i^4 = i^2 i^2 \rightarrow i^4 = (-1)(-1) \rightarrow i^4 = 1$. Why does this matter? Well, if we had to simplify something like i^8, we can conclude that $(i^8) = (i^4)^2 = (1)^2 = 1$. Basically, if the exponent to your imaginary number is a multiple of 4, you know it simplifies to be 1! Let's take a look at this problem: $3i^{87}$ Well, 87 is NOT a multiple of 4, but 84 is: (*Note: you have to use key exponent rules!*) Look: $3i^{87} = 3i^{84}(i^3)$. Now, we know that $i^{84} = 1$ and we have: $3i^{87} = 3(1)(i^3)$. Remember: $i^2 = -1$. So, $3i^{87} = 3(1)(i^3) = 3i^2(i) = -3i$.

7) Our first step in solving for x here is to bring all terms to one side of the equal sign:

$x^2 + 3x - 7 = 3 \rightarrow x^2 + 3x - 10 = 0$. Now, look to see if you can factor. In this case, are there any two numbers such that their product is -10 and their sum is 3? I think I know of two! 5 and -2! So, we can factor:

$x^2 + 3x - 10 = 0 \rightarrow (x + 5)(x - 2) = 0$. This tells us $x + 5 = 0$ or $x - 2 = 0$. Now, we'll solve for x in both equations to find our solutions: $x + 5 = 0 \rightarrow x = -5$ and $x - 2 = 0 \rightarrow x = 2$.

8) Let's follow the same steps as those in question #7's explanation: we begin already with all terms on one side of the equal sign: $9x^2 - 25 = 0$. So, now we want to see if we can factor. We notice that this problem presents the difference between two perfect squares! Look: $9x^2 - 25 = (3x)^2 - (5)^2$. What does this mean? Well, we know that there is a way in which we factor the difference between perfect squares:

$a^2 - b^2 = (a - b)(a + b)$. In this problem, $a = 3x$ and $b = 5$. So,

$9x^2 - 25 = 0 \rightarrow (3x)^2 - (5)^2 = 0 \rightarrow (3x - 5)(3x + 5) = 0$. This tells us that $3x - 5 = 0$ or $3x + 5 = 0$.

Let's solve both equations to find our values of x: $3x - 5 = 0 \rightarrow x = \dfrac{5}{3}$ and $3x + 5 = 0 \rightarrow x = \dfrac{-5}{3}$.

9) We begin by bringing all terms to one side of the equation: $5x^2 = 7x - 6 \rightarrow 5x^2 - 7x + 6 = 0$. Now, we'll see if we can factor: $5x^2 - 7x + 6 = (5x - ?)(x - ?)$. Are there any numbers that make sense? NO!

Instead, let's use the **quadratic formula** to find the possible values of x so that:: $x_1 = \dfrac{-b + \sqrt{b^2 - 4ac}}{2a}$ and

$x_2 = \dfrac{-b - \sqrt{b^2 - 4ac}}{2a}$ for a quadratic equation in the form of : $ax^2 + bx + c = 0$. For this problem, $a = 5$,

$b = -7$ and $c = 6$. We solve by plugging in the appropriate values and simplifying:

$x = \dfrac{7 \pm \sqrt{-71}}{10} \rightarrow x = \dfrac{7 \pm i\sqrt{71}}{10}$

10) We begin by bringing all terms to one side of the equation: $2x^2 - 9x = -\dfrac{1}{2} - 6x \rightarrow 2x^2 - 3x + \dfrac{1}{2} = 0$.

After trying to factor, we see that our best bet is to use the **quadratic formula** as illustrated in question #9:

Here, $a = 2$, $b = -3$ and $c = \dfrac{1}{2}$. We solve by plugging in the appropriate values and simplifying:

$x_1 = \dfrac{-(-3) + \sqrt{(-3)^2 - 4(2)(\frac{1}{2})}}{4} \rightarrow \dfrac{3 + \sqrt{5}}{4}$, $x_2 = \dfrac{-(-3) - \sqrt{(-3)^2 - 4(2)(\frac{1}{2})}}{4} \rightarrow \dfrac{3 - \sqrt{5}}{4}$

11) We begin by bringing all terms to one side of the equation, making sure that the x^2 value has a positive coefficient! $-x^2 = -2x - 3 \rightarrow x^2 - 2x - 3 = 0$. Let's try to factor: Are there any two terms such that their product is -3 and their sum is -2? I know! -1 and $+3$! So, $x^2 - 2x - 3 = 0 \rightarrow (x - 3)(x + 1) = 0$. This tells us that $x - 3 = 0$ or $x + 1 = 0$. Let's solve both equations to find our values of x:

$x - 3 = 0 \rightarrow x = 3$ and $x + 1 = 0 \rightarrow x = -1$.

12) Let's begin by bringing all terms to one side of the equation, making sure that the x^2 value has a positive coefficient! $6x + 13x^2 - 12 = 3x - 15 + 12x^2 \rightarrow x^2 + 3x + 3 = 0$. Let's try to factor: are there any two integers such that their product is 3 and their sum is 3? No! We see that our best bet is to use the **quadratic formula** as illustrated in questions #9 and #10: Here, $a = 1$, $b = 3$ and $c = 3$. We solve by plugging in the appropriate

values and simplifying: $x_1 = \dfrac{-(3) + \sqrt{(3)^2 - 4(1)(3)}}{2} \rightarrow \dfrac{-3 + i\sqrt{3}}{2}$ $x_2 = \dfrac{-(3) - \sqrt{(3)^2 - 4(1)(3)}}{2} \rightarrow \dfrac{-3 - i\sqrt{3}}{2}$

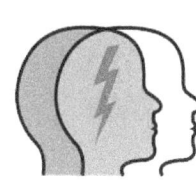

Directions: Answer the following questions by showing all your work. When possible, avoid using your calculator.

Simplify:

1) $\dfrac{7}{18} - \dfrac{10}{6}$

2) $\dfrac{x}{h} - \dfrac{6y}{k}$

3) $\dfrac{5}{a} - \dfrac{6y}{a+6}$

Solve for x:

4) $\dfrac{x+6}{7} = \dfrac{3x}{9}$

5) $\dfrac{9-2}{x+1} = 4y+2$

6) $\dfrac{2x}{d} - \dfrac{r}{w} = c$

Simplify:

7) $\sqrt{18}$

8) $7 + 2\sqrt{3} + 5\sqrt{2} + 6\sqrt{3} - \sqrt{2}$

9) $4\sqrt{75} - 2\sqrt{147} + \sqrt{3}$

10) $\dfrac{4 - \sqrt{8}}{2}$

11) $(2\sqrt{5})^2$

12) $\sqrt{2} \cdot \sqrt{18}$

13) $\dfrac{a\sqrt{a^2}}{\sqrt{a}}$

Rationalize the denominator:

14) $\dfrac{1}{3 + \sqrt{2}}$

15) $\dfrac{\sqrt{2}+1}{\sqrt{2}-1}$

16) $\dfrac{7}{3\sqrt{5} - \sqrt{3}}$

17) $\dfrac{-1}{\sqrt[3]{ab}}$

18) $\sqrt[3]{\dfrac{3a^4}{7b^2}}$

1) **Remember:** we can only add or subtract fractions if they have the same denominator (bottom number). If they do not, you must establish a common denominator by calculating the Least Common Multiple between the denominators of the given fractions. For example, the denominators of the fractions in this problem are 18 and 6. We know that least common multiple is 18. Once we've found what the common denominator should be, we want to adjust both fractions so they BOTH include this denominator: In this case, we need to change

$\frac{10}{6}$ only: $\frac{10}{6}\left(\frac{3}{3}\right) = \frac{30}{18}$. Why are we able to multiply $\frac{10}{6}$ by $\frac{3}{3}$? Because multiplying by $\frac{3}{3}$ is the same as

multiplying by 1! Now, we are ready to subtract and simplify: $\frac{7}{18} - \frac{30}{18} = \frac{-23}{18}$. (**Note:** when adding or

subtracting fractions we add or subtract across the numerators (top numbers) only)

2) (see explanation in question #1) The common denominator here is hk. We have to adjust both fractions:

$\frac{x}{h}\left(\frac{k}{k}\right) = \frac{xk}{hk}$ and $\frac{6y}{k}\left(\frac{h}{h}\right) = \frac{6yh}{hk}$. Now, subtract and simplify: $\frac{xk}{hk} - \frac{6yh}{hk} = \frac{xk - 6yh}{hk}$.

3) (see explanation in question #1) The common denominator here is $a(a+6)$. So, we have to adjust both

fractions: $\frac{5}{a}\left(\frac{a+6}{a+6}\right) = \frac{5a+30}{a^2+6a}$ and $\frac{6y}{a+6}\left(\frac{a}{a}\right) = \frac{6ay}{a^2+6a}$. Now, subtract and simplify:

$\frac{5}{a} - \frac{6y}{a+6} \rightarrow \frac{5a+30}{a^2+6a} - \frac{6ay}{a^2+6a} = \frac{5a-6ay+30}{a^2+6a}$

4) Let's solve for x by cross-multiplying (multiplying the numerator of one fraction by the denominator of the

other fraction: $\frac{x+6}{7} = \frac{3x}{9} \rightarrow (3x)(7) = (x+6)(9) \rightarrow 21x = 9x + 54 \rightarrow x = 4.5$

5) We begin by multiplying both sides of the equation by $(x+1)$ and then solve for x:

$(x+1)\frac{9-2}{x+1} = (4y+2)(x+1) \rightarrow 7 = (4y+2)(x+1) \rightarrow \frac{7}{4y+2} = x+1 \rightarrow x = \frac{7}{4y+2} - 1$

6) Let's begin by getting rid of the denominators: to do so, we'll multiply both sides of the equation by dw

and then solve for x: $(dw)\left(\dfrac{2x}{d} - \dfrac{r}{w}\right) = c(dw) \rightarrow 2wx - rd = cdw \rightarrow 2wx = cdw + rd \rightarrow x = \dfrac{cdw + rd}{2w}$

7) To simplify this term, we begin by finding a perfect square factor underneath the radical. In this case, it is 9:

$\sqrt{18} = \sqrt{9 \cdot 2}$. Now, convert this into the product of two radicals and simplify: $\sqrt{9 \cdot 2} = \sqrt{9}\sqrt{2} = 3\sqrt{2}$.

8) Let's simplify by combining 'like terms': $7 + 2\sqrt{3} + 5\sqrt{2} + 6\sqrt{3} - \sqrt{2} = 8\sqrt{3} + 4\sqrt{2} + 7$

9) First, let's simplify the terms of the expression: $4\sqrt{75} = 4\sqrt{25}\sqrt{3} = 20\sqrt{3}$, $2\sqrt{147} = 2\sqrt{49}\sqrt{3} = 14\sqrt{3}$

and $\sqrt{3} = \sqrt{3}$. So, $4\sqrt{75} - 2\sqrt{147} + \sqrt{3} = 20\sqrt{3} - 14\sqrt{3} + \sqrt{3} = 7\sqrt{3}$.

10) First, we simplify the radical $\dfrac{4 - \sqrt{8}}{2} = \dfrac{4 - 2\sqrt{2}}{2}$. Then, divide by 2: $\dfrac{4 - \sqrt{8}}{2} = \dfrac{\cancel{4} - \cancel{2}\sqrt{2}}{\cancel{2}} = 2 - \sqrt{2}$

11) Use the law of exponents $(xy)^a = x^a y^a$ to solve: $(2\sqrt{5})^2 = 4 \cdot 5 = 20$

12) Let's simply multiply the numbers underneath the radical and simplify: $\sqrt{18}\sqrt{2} = \sqrt{36} = 6$

13) We use the property that $\dfrac{\sqrt{x}}{\sqrt{y}} = \sqrt{\dfrac{x}{y}}$ and the law of exponents that $\dfrac{x^n}{x^p} = x^{n-p}$: $\dfrac{a\sqrt{a^2}}{\sqrt{a}} = a\left(\sqrt{\dfrac{a^2}{a}}\right) = a\sqrt{a}$

14) We rationalize the denominator of a fraction by multiplying the numerator and the denominator by the denominator's conjugate. What's a conjugate? In general, the conjugate of $a + b$ is $a - b$. In this case, the

denominator's conjugate is $3 - \sqrt{2}$. Let's see what we get: $\dfrac{1}{3 + \sqrt{2}}\left(\dfrac{3 - \sqrt{2}}{3 - \sqrt{2}}\right) = \dfrac{3 - \sqrt{2}}{9 - 2} = \dfrac{3 - \sqrt{2}}{7}$

15) Rationalize by multiplying the top and bottom of the fraction by the denominator's conjugate:

$\dfrac{\sqrt{2} + 1}{\sqrt{2} - 1}\left(\dfrac{\sqrt{2} + 1}{\sqrt{2} + 1}\right) = \dfrac{2 + 1 + 2\sqrt{2}}{1} = 3 + 2\sqrt{2}$

145

16) (see questions #14-15) $\dfrac{7}{3\sqrt{5}-\sqrt{3}}\left(\dfrac{3\sqrt{5}+\sqrt{3}}{3\sqrt{5}+\sqrt{3}}\right)=\dfrac{21\sqrt{5}+7\sqrt{3}}{45-3}=\dfrac{21\sqrt{5}+7\sqrt{3}}{42}\rightarrow\dfrac{3\sqrt{5}+\sqrt{3}}{6}$

17) To rationalize the denominator of this term, we want to start by multiplying the top and bottom of the fraction

by $\sqrt[3]{a^2b^2}$: $\dfrac{-1}{\sqrt[3]{ab}}\left(\dfrac{\sqrt[3]{a^2b^2}}{\sqrt[3]{a^2b^2}}\right)=\dfrac{-\sqrt[3]{a^2b^2}}{ab}$

18) Begin by using the property that $\sqrt{\dfrac{x}{y}}=\dfrac{\sqrt{x}}{\sqrt{y}}$ and then rationalize the denominator by multiplying the top

and bottom of the fraction by $\sqrt[3]{7^2b}$: $\sqrt[3]{\dfrac{3a^4}{7b^2}}=\dfrac{\sqrt[3]{3a^4}}{\sqrt[3]{7b^2}}\left(\dfrac{\sqrt[3]{7^2b}}{\sqrt[3]{7^2b}}\right)=\dfrac{\sqrt[3]{147a^4b}}{7b}=\dfrac{a\sqrt[3]{147ab}}{7b}$

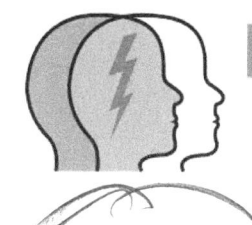

Directions: Answer the following questions by showing all your work. When possible, avoid using your calculator.

Simplify:

1) 23^1

2) $64^{\frac{1}{2}}$

3) 17^0

4) $\dfrac{18a^2b^5c^2}{a^{10}6c^{-5}b^4}$

5) $\dfrac{(6^2x^3)^6}{x^2x^3}$

6) $7y^2 + 6y - 3y^2$

7) $(z^{-3} \cdot 3z^6)^{\frac{1}{2}}$

8) $(x^4)(y^3)$

9) $\left[(3x^4y^7z^{12})^5(-5x^9y^3z^{11})^2\right]^0$

Find x:

10) $\log_x 27 = 3$

11) $\log_{10} 1000 = x$

12) $\log_4 x = 3$

13) $10^{x+5} - 8 = 60$

14) $4\log_x 2 = 1$

15) $4^{x+3} = (64)^x$

Expand:

16) $\log_{10}\left(\dfrac{x}{3y}\right)$

17) $\log_4 4\left(\dfrac{a}{3}\right)^2$

Simplify:

18) $\log_{10} 7 + \log_{10} 3$

19) $\log_3 3^9$

1) Any number raised to the first power is, in fact, the number itself: $23^1 = 23$

2) Any number raised to the one-half power is the square root of that number: $64^{\frac{1}{2}} = \sqrt{64} = \pm 8$

3) Don't be fooled! ANY number raised to the zero power is 1! $17^0 = 1$

4) Let's work backwards: Let's factor the fraction into smaller fractions comprised of like terms. Then, simplify the smaller fractions and multiply to create one term. We use the exponent rule: $\dfrac{x^n}{x^p} = x^{n-p}$

$$\frac{18a^2b^5c^2}{6a^{10}b^4c^{-5}} = \left(\frac{18}{6}\right)\left(\frac{a^2}{a^{10}}\right)\left(\frac{b^5}{b^4}\right)\left(\frac{c^2}{c^{-5}}\right) = 3(a^{2-10})(b^{5-4})(c^{2-(-5)}) = 3a^{-8}bc^7 \to \frac{3bc^7}{a^8}$$

5) For this, we use the exponent rules: $(x^a y^b)^n = x^{an}y^{bn}$ and $x^a x^b = x^{a+b}$. We also will follow the same solving process as that of the previous question: $\dfrac{(6^2 x^3)^6}{x^2 x^3} = \dfrac{6^{12} x^{18}}{x^5} = 2176782336\left(\dfrac{x^{18}}{x^5}\right) = 2176782336x^{13}$

6) This is simple: simplify by adding 'like-terms': $7y^2 + 6y - 3y^2 = 4y^2 + 6y$

7) We use the exponent rules described in question #5: $(z^{-3} \cdot 3z^6)^{\frac{1}{2}} = (3z^3)^{\frac{1}{2}} = 3^{\frac{1}{2}} z^{\frac{3}{2}} \to \sqrt{3} \cdot \sqrt{z^3} = z\sqrt{3z}$

8) Don't be tricked! We can't really simplify this any further because we aren't dealing with the SAME base. Let's just multiply and get rid of the parentheses: $(x^4)(y^3) = x^4 y^3$

9) *Remember!* Any number raised to the zero power is 1! $\left\lceil \left[(3x^4 y^7 z^{12})^5 (-5x^9 y^3 z^{11})^2 \right] \right\rceil^0 = 1$

10) For this, we are going to use the common definition of a logarithm: if $n^a = p \to \log_n p = a$. So:
$\log_x 27 = 3 \to x^3 = 27 \to x = \sqrt[3]{27} \to x = 3$

11) Use the definition and solve for x: $\log_{10} 1000 = x \to 10^x = 1000 \to x = 3$.

12) $\log_4 x = 3 \to 4^3 = x \to 64 = x$

13) Let's begin by isolating the variable, x. Then, convert the equation into a logarithm to solve for x:
$10^{x+5} - 8 = 60 \to 10^{x+5} = 68 \to \log_{10} 68 = x + 5 \to x = \log_{10} 68 - 5 \to x = -3.17$ (log on calc).

14) Using the rules of logs, we know that $4 \log_x 2 = 1 \rightarrow \log_x 2^4 = 1$. Therefore,

$\log_x 2^4 = 1 \rightarrow \log_x 16 = 1 \rightarrow x^1 = 16 \rightarrow x = 16$

15) We'll first change $(64)^x$ so it has a base of 4: $(64)^x = (4^3)^x = 4^{3x}$. So, $4^{x+3} = 4^{3x}$. Therefore,

$x + 3 = 3x \rightarrow x = \frac{3}{2}$.

16) To expand this term, we'll use the following rule: $\log \frac{a}{b} = \log a - \log b$. (***remember*** rules of exponents

are similar to those of logarithms): So, $\log_{10}\left(\frac{x}{3y}\right) = \log_{10} x - \log_{10} 3y = \log_{10} x - (\log_{10} 3 + \log_{10} y)$

17) To expand this term, we'll use the rule described in question #16 and $\log ab = \log a + \log b$ So,

$\log_4 4 \left(\frac{a}{3}\right)^2 = \log_4 4 + \log_4 \frac{a}{3} + \log_4 \frac{a}{3} = 1 + \log_4 a - \log_4 3 + \log_4 a - \log_4 3 = 1 + 2\log_4 a - 2\log_4 3$

18) Let's simplify by using the rule described in question #17: $\log_{10} 7 + \log_{10} 3 = \log_{10} 21 = 1.32$

19) We'll use the definition of logarithms to simplify the term: $\log_3 3^9 = x \rightarrow 3^x = 3^9 \rightarrow x = 9$

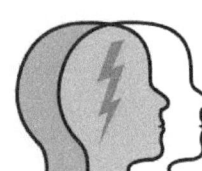

Directions: Answer the following questions by showing all your work. When possible, avoid using your calculator.

<u>Solve</u> the systems of equations numerically & indicate their **graphical** relationships:

1) $5x + 3y = 7$
$-5y + 23 = -3x$

2) $2x - 4y = -18$
$3x + 9y = 48$

3) $3x + 4y = 7$
$9x + 12 = -12y$

4) $y = 36 - 9x$
$3x + \frac{1}{3}y = 12$

5) $3x - y + 2 = -3z$
$-2x + y - 4z = 10$
$x - 4y + z = 2$

Solve:

6) The ratio of 9 more than x to x equals 4. Find x. _____

7) Seven less than the quotient of x and 3 is zero. Find x. _____

8) The length of a football field is 30 yards more than its width, w. Express the length of the field in terms of w. _____

9) What percent of 20 is 30? _____

10) 45% of what is 9? _____

11) What is 35% of 80? _____

12) The sum of the digits of a two–digit number is 7. When the digits are reversed, the number is increased by 27. Find the number. _____

13) A wallet contains 13 coins of only nickels and pennies. They total $.37. How many of each coin are there in the wallet? _____

Solving Systems of Equations: We solve systems of equations to find out where the given equations, graphically, intersect. We can do this in two different ways: (1) using strictly substitution or (2) elimination. We'll employ one of these methods in solving the systems of equations below. This decision is based on which method is most appropriate.

1) We'll solve this system of equations using substitutions. First, we'll solve for x in the first equation: $5x + 3y = 7 \rightarrow x = \dfrac{7 - 3y}{5}$. Now we'll substitute this into the second equation and solve for y:

$-5y + 23 = -3\left(\dfrac{7 - 3y}{5}\right) \rightarrow -5y + 23 = \left(\dfrac{-21 + 9y}{5}\right) \rightarrow -25y + 115 = 21 + 9y \rightarrow y = 4$. We solve for x by

substituting $y = 4$ into $x = \dfrac{7 - 3y}{5}$. Therefore, $x = \dfrac{7 - 3(4)}{5} \rightarrow x = -1$. So, graphically, these two lines

intersect at the point: $(-1, 4)$.

2) We'll solve this system of equations using elimination. To do this, we want to manipulate what we have so that the 'x' coefficient is the SAME in both equations. First, we notice that you can divide the first equation

through by 2: $\dfrac{2x - 4y = -18}{2} \rightarrow x - 2y = -9$. (now our 'x' coefficient is 1) We also notice that you can

divide the second equation through by 3: $\dfrac{3x + 9y = 48}{3} \rightarrow x + 3y = 16$ (now our 'x' coefficient is 1). Now,

let's eliminate the 'x' variable by SUBTRACTING the equations from one another: $\begin{array}{r} x - 2y = -9 \\ -x - 3y = -16 \\ \hline -5y = -25 \end{array}$. Now, we

solve for y: $-5y = -25 \rightarrow y = 5$. We solve for x by substituting $y = 5$ into $x + 3y = 16$. Therefore,

$x + 3(5) = 16 \rightarrow x = 1$. So, graphically, these two lines intersect at the point: $(1, 5)$.

3) First, let's change the second equation so it looks like the first one: $9x + 12 = -12y \rightarrow 9x + 12y = -12$.

We'll solve this equation using elimination. Let's multiply the first equation by -3 so we can eliminate the 'x'

terms in the equations when we combine them: $(-3)3x + (-3)4y = (-3)7 \rightarrow -9x - 12y = -21$. Let's

combine this with the second equation: $\begin{array}{r} -9x - 12y = -21 \\ 9x + 12y = -12 \\ \hline 0 = -33 \end{array}$, which obviously, cannot be true. So, we have *no*

solution. This means that *these lines DO NOT intersect.* Graphically, they are *parallel lines.*

151

4) We'll definitely use substitution for this system of equations! Substitute $y = 36 - 9x$ in for y in the second equation and solve for x: $3x + \dfrac{1}{3}(36 - 9x) = 12 \rightarrow 3x + 12 - 3x = 12 \rightarrow 0 = 0$. Therefore, no matter what x or y values you plug into these equation they will still be the same! Try it! Therefore, *these lines intersect in infinitely many points*. In other words, graphically they describe the same line!

5) We'll solve this system using elimination and substitution. First, we'll change the first equation to look like the other two: $3x - y + 2 = -3z \rightarrow 3x - y + 3z = -2$. We notice that this equation has a negative y; while the second equation has a positive x. So, we can eliminate the y variable from these equations by ADDING them: $\begin{array}{l} 3x - y + 3z = -2 \\ \underline{-2x + y - 4z = 10} \\ \quad x - z = 8 \end{array}$. Now, we solve for z: $z = x - 8$. Let's substitute this in for z in the first equation and simplify: $3x - y + 3(x - 8) = -2 \rightarrow 6x - y = 22$ *and* third equation and simplify:

$x - 4y + (x - 8) = 2 \rightarrow 2x - 4y = 10$. We'll multiply $2x - 4y = 10$ by -3 so we have: $(-3)2x - (-3)4y = (-3)10 \rightarrow -6x + 12y = -30$. Now, we can combine it with $6x - y = 22$:

$\begin{array}{l} 6x - y = 22 \\ \underline{-6x + 12y = -30} \\ \quad 11y = -8 \end{array}$, which tells us that $y = -8/11$. Let's substitute $-8/11$!78*!?h $2x - 4y = 10$ to find the

value of x: $2x - 4(-8/11) = 10 \rightarrow x = 39/11$. And, $z = x - 8 \rightarrow z = 39/11 - 8 \rightarrow z = -49/11$. Therefore,

graphically, the lines that represent these equations intersect at the point: $(x, y, z) = (39/11, -8/11, -49/11)$.

6) For this question, we want to translate words into numbers! Let's split it up into parts: The ratio $\dfrac{?}{?}$ of 9 more than a number x: $\dfrac{x + 9}{?}$ to x: $\dfrac{x + 9}{x}$ is 4: $\dfrac{x + 9}{x} = 4$. Now, we'll solve for x by using cross-multiplication: $\dfrac{x + 9}{x} = 4 \rightarrow 4x = x + 9 \rightarrow 3x = 9 \rightarrow x = 3$.

7) For this question, we want to translate words into numbers! Let's split it up into parts: seven less -7 than a quotient $\dfrac{?}{?} - 7$ of x and 3: $\dfrac{x}{3} - 7$ is 0: $\dfrac{x}{3} - 7 = 0$. Now, solve for x : $\dfrac{x}{3} - 7 = 0 \rightarrow \dfrac{x}{3} = 7 \rightarrow x = 21$.

8) Allow the length of the football field = l and the width of the football field = w. Employ the same methods of 'translation' as questions #6 and #7: $l = 30 + w$. (length expressed in terms of width, w)

9) First, taking $n\%$ of a number is the same as multiplying that number by $\dfrac{n}{100}$. Therefore, $20\left(\dfrac{n}{100}\right) = 30$.

Solve for n: $20\left(\dfrac{n}{100}\right) = 30 \rightarrow n = 150\%$.

10) Remember that taking 45% of a number is the same as multiplying that number by $\dfrac{45}{100}$. We are told that 45% of a number, n is 9. We'll set up the equation and solve for n: $\left(\dfrac{45}{100}\right)n = 9 \rightarrow n = 20$.

11) (see question #9) 35% of 80: $80\left(\dfrac{35}{100}\right) = 28$

12) Again, we work through this problem by translation: If the sum of the digits of a two-digit number is 7, let's allow the 'tens' digit of the number equal x and the 'ones' digit of this number equal y, so $x + y = 7$, When the digits are reversed, the number is increased by 27: $10x + y + 27 = 10y + x$. Since $x = 7 - y$ we can substitute $7 - y$ in for x in our second equation and solve for y:

$10(7 - y) + y + 27 = 10y + (7 - y) \rightarrow 90 = 18y \rightarrow y = 5$. Therefore, $x = 7 - y \rightarrow x = 2$. So, our original number is $\underline{x}\,\underline{y} = \underline{2}\,\underline{5}$.

(**Note:** it is easy to take an alternate approach- conceptually, there are only a limited number of integer pairs that add to 7. They are $(0,7), (1,6), (5,2), (3,4)$. Guess and check to see which ones work.)

13) Allow $13 = n + p$, where n=number of nickels and p=number of pennies. Because a nickel is worth \$.05 and a penny is worth \$.01, we know that $.05n + .01p = .37$. But there are $n = 13 - p$ nickels. Substitute for n and solve for p: $.05(13 - p) + .01p = .37 \rightarrow .65 - .05p + .01p = .37 \rightarrow -.04p = -.28 \rightarrow p = 7$. Plug this into $n = 13 - p$ and solve for n: $n = 13 - 7 \rightarrow n = 6$. Therefore, there are seven pennies and six nickels!

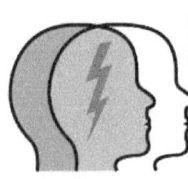

Directions: Answer the following questions by showing all your work. Note that figures are **not** drawn to scale. Be sure to use correct units in your answers.

(1) Given $f(x)= x^2-5$, find: (a) $f(-1)$ ___-4___ (b) $f(y+2)$ ___y^2+4y-1___ (c) $f^{-1}(3)$ ___

$(-1)^2 - 5$ $(y+2)^2 - 5$

$1 + 5$ $(y+2)(y+2) - 5$

-4

$y^2 + 2y + 2y + 4$

$y^2 + 4y + 4 - 5$

$y^2 + 4y - 1$

(2) If $g(x)=51+2x$ and $h(x)=x^{1/3}+2$, find $g(h(-64))$ ___57___

$-64^{\frac{1}{3}} + 2$

$-4 + 2 = -2$

$51 + 2(-2)$

$51 - 4$

(3) If $f(x)=x^3-64x$ find the values at which the functions graph passes through the x-asis. ___

x-int:

$0 = x^3 - 64x^1$

$0 = x(x^2 - 64)$

$(x-8)(x+8)$

$x = 0$

$x = \pm 8$

(4) If $k(x)=\sin x$ and $0 < x < 360°$, find the values of x when $k(x)= \dfrac{-1}{2}$

(5) To the right is the graph of the function $f(x)=y$.
For how many values of x does $f(x)=3$? ___

154

(1) Given $f(x) = x^2 - 5$, we find $f(-1)$ by allowing $x = -1$ for $f(x) = x^2 - 5$. So, $f(-1) = (-1)^2 - 5 \rightarrow f(-1) = -4$. Similarly, we find $f(y + 2)$ by plugging in $y + 2$ for x in $f(x) = x^2 - 5$. So, $f(y + 2) = (y + 2)^2 - 5 \rightarrow f(y + 2) = y^2 + 4y - 1$. Now, to find $f^{-1}(3)$, we first have to find $f^{-1}(x)$, the inverse of the function $f(x) = x^2 - 5$. We start by substituting y in for $f(x)$: $y = x^2 - 5$. Now we switch the y and x values in the equation and solve for y: $x = y^2 - 5 \rightarrow y = \sqrt{x} + 5$. Now, we substitute $f^{-1}(x)$ for y and you have our inverse function $f^{-1}(x) = \sqrt{x} + 5$. So, to solve for $f^{-1}(3)$, we allow $x = 3$ in for $f^{-1}(x)$: $f^{-1}(3) = \sqrt{3} + 5 \rightarrow f^{-1}(3) = 2\sqrt{2}$.

(2) To find the value of $g(h(-64))$ we have to work from the inside, OUT! So, first we want to find the value of $h(-64)$. Like in question #1, we allow $x = -64$ for $h(x) = x^{\frac{1}{3}} + 2$: $h(-64) = (-64)^{\frac{1}{3}} + 2 = -2$. Now, we have simplified the problem to $g(-2)$ and find this value by allowing $x = -2$ for $g(x) = 51 + 2x$: $g(-2) = 51 + 2(-2) = 47$. In conclusion: $g(h(-64)) = 47$.

(3) The values at which the function's graph passes through the x-axis are also known as the **x-intercepts** of the function. We find the x-intercepts of $f(x)$ by setting the function, $f(x) = x^3 - 64x$ to zero and solving for x. (because $f(x) = y = 0$ on the x-axis) So, $0 = x^3 - 64x \rightarrow 0 = x(x^2 - 64)$. This tells us that $x = 0$ or $x^2 - 64 = 0 \rightarrow x = \sqrt{64} \rightarrow x = \pm 8$. Therefore, the graph passes through the x-axis at: $x = 8$, $x = -8$ and $x = 0$.

(4) We are given $k(x) = \sin x$, $0 < x < 360°$ (radians: $0 < x < 2\pi$). Now, we find the value of x when $k(x) = -\frac{1}{2}$ by allowing $\sin x = -\frac{1}{2}$ and finding an angle, x, $0 < x < 2\pi$ that satisfies the equation. Knowledge of the Unit Circle would be very useful here! We want to know: for which angles(x) is the sine (or y-coordinate of its terminal point) equal to $-\frac{1}{2}$? We know that $\sin \frac{7\pi}{6} = -\frac{1}{2}$ and $\sin \frac{11\pi}{6} = -\frac{1}{2}$. Therefore, $x = \frac{7\pi}{6}$ and $x = \frac{11\pi}{6}$. Or, in degrees, $x = 210°$ and $x = 330°$

(5) In the graph of the function $f(x) = y$ to the right, $f(x) = 3$ when $y = 3$. We can see that $y = 3$ for *three* values of x as marked on the graph.

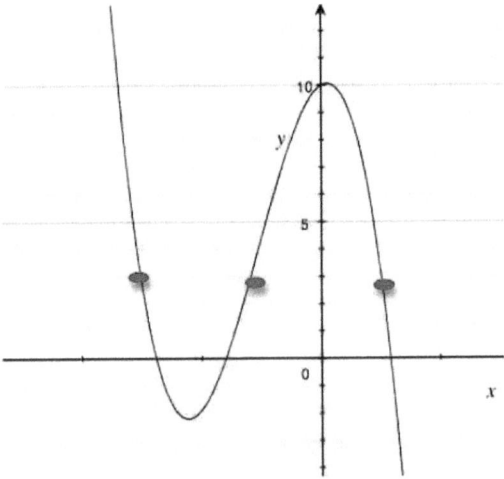

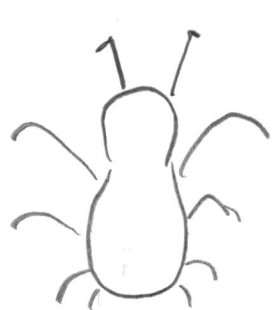

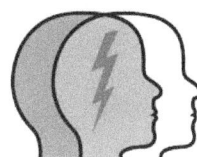

Directions: Answer the following questions by showing all your work.

Note that figures are **not** drawn to scale.

Be sure to use correct units in your answers.

(1) If $j*k = (j-k)^2$, what is the value of $-2*-3$? _____

(2) If $a!b$ represents the set of prime numbers between a and b, inclusive, what is $5!55$? _____

(3) Ken drives at 40 miles per hour. At this rate, how many minutes will it take him to drive 50 miles? _____

(4) Jordan walks to school at a rate of 4 miles per hour and returns home at 7 miles per hour. If the total trip took 5.5 hours, how far from the school is Jordan's home? _____

(5) If Jethro can mow h lawns in k hours, how many lawns can he mow in m hours? _____

(6) Micaiah currently has test scores of 83, 75 and 88. What does she need to get on her last exam (double-graded) to finish the course with an 87 average?_____

(7) How many distinct 5-people teams can be made of a group of 9 people?

(8) How many outfit combinations can Derek make if he has 8 ties, 7 shirts, and 4 pairs of pants? _____

(1) We are given $j * k = (j - k)^2$. So, to solve $-2 * -3$, we substitute: $j = -2$ and $k = -3$. Therefore, $-2 * -3 = (-2 - (-3))^2 = 1$.

(2) We are given $a!b$ represents the set of prime numbers between a and b, inclusive. So, to solve $5!55$ we substitute: $a = 5$ and $b = 55$. (Please review your prime numbers between 1 and 100!) Therefore, $5!55 = \{5, 7, 11, 13, 17, 19, 23, 29, 31, 37, 41, 43, 47, 53\}$

(3) We are given that Ken drives at 40 miles per hour. We can set up a fraction to represent this rate: $\dfrac{40mi}{1hr}$.

To find how long it takes him to drive 50 miles we set up a proportion and solve for the missing value, x:

$\dfrac{40mi}{1hr} = \dfrac{50mi}{xhr} \rightarrow 40x = 50 \rightarrow x = 1.25hr$. This is great! But, we need our answer in **minutes**! Well, how

many minutes are in $1.25hr$? We'll set up another proportion and solve for our missing value, x:

$\dfrac{60min}{1hr} = \dfrac{xmin}{1.25hr} \rightarrow x = 60(1.25) = 75min$

(4) With the information given, let's represent both of Jordan's trips (using $D = rate \times time$) with the equations:

(a) To school: $D = 4t_1$ and (b) From school $D = 7t_2$ where D = distance from her home to school and t_1 = the

time it took to walk to school and t_2 = time it took to walk from school. But, since D, the distance to and from

school remains constant, we can say: $4t_1 = 7t_2$. Now, we are given that the total trip took 5.5 hours, meaning

that $t_1 + t_2 = 5.5 \rightarrow t_1 = 5.5 - t_2$. Now we can substitute $5.5 - t_2$ for t_1 in $4t_1 = 7t_2$ and solve for t_2: $4(5.5 - t_2) = 7t_2 \rightarrow 22 = 11t_2 \rightarrow t_2 = 2hr$. We can find the distance, D by plugging this value of into our second

equation, $D = 7t_2$. So, distance, $D = 7(2) = 14 \; miles$

(5) We are given that Jethro can mow h lawns in k hours and want to know how many lawns he can mow in

m hours, we set up a proportion and solve for our missing value, x, $\dfrac{lawns}{hours} \rightarrow \dfrac{h}{k} = \dfrac{x}{m} \rightarrow xk = mh \rightarrow x = \dfrac{mh}{k}$.

(6) The average of four test scores in which one is double graded is given by: $A_{tests} = \dfrac{T_1 + T_2 + T_3 + 2T_4}{5}$. We

multiply T_4 by 2 and divide by 5 when we take the average because T_4 is double-graded- technically counted

for two times! In Micaiah's case, allow $T_1 = 83$, $T_2 = 75$ $T_3 = 88$ and $A_{tests} = 87$. We want to substitute

these values into the equation given and solve for T_4: $87 = \dfrac{83 + 75 + 88 + 2T_4}{5} \rightarrow 2T_4 = 189 \rightarrow T_4 = 94.5$

(7) To solve this problem, we have to have knowledge of **Combinatorics.** This problem—*how many 5-people*

teams can you make out of 9 people—asks for combinations, without allowing for repetition. The number of k-

combinations made out of a set of n-elements is given by the formula: $_nC_k = \dfrac{n!}{k!(n-k)!}$. In this case, we

want the number of 5-people combinations made out of a set of 9 people. Let's substitute the appropriate

values and simplify to find the amount of teams we can make: $_9C_5 = \dfrac{9!}{5!(9-5)!} = 126$.

(8) To find how many outfit combinations Derek can make if he has 8 ties, 7 shirts, and 4 pairs of pants, we

simply want to multiply the number of items in the three different groupings: $8 \times 7 \times 4 = 224$ outfit

combinations.

(8) In a school assembly there are 350 students, all freshmen and sophomores. There are 90 more freshmen than sophomores. Three-fifths of the sophomores are girls. How many sophomore boys are there in the assembly? _____

(9) If set A is the set of all prime numbers and set B is the set of all multiples of 7, how many terms are in the intersection of A and B? _____

(10) If $a < b < 0 < c$, which of the following must be true? _____

    I. $c^b < c$

    II. $b - a > o$

    III. $b^2 > a$

Directions: Answer the following questions by showing all your work. Note that figures are **not** drawn to scale. Be sure to use correct units in your answers.

(1) If the remainder when $j+3$ is divided by 5 is 3, what is a possible value of j^2? _____

(2) Two-digit numbers AA and BB add to a three-digit number as shown.

$$AA$$
$$+BB$$
$$\overline{AAC}$$

What is the value of C? _____

(3) This year Cory earned $85,000. Last year he made $60,000. If his income increases next year by the same percent, how much will he make next year? _____

(4) What is the 249th term of the following sequence? 8,3,-1,2,4,5,8,3,-1,2,4,5... _____

(5) In a lot of used cars that are numbered from 1 to 125, James buys the cars numbered 18 to 33. Five-eights of these cars are grey while the rest are magenta. How many of these cars are magenta? _____

(6) The average of four consecutive odd numbers is 32. What is the product of the least and the greatest values in this set of integers? _____

(7) What is the sum of the greatest prime number less that 93 and the least prime number greater than 50? _____

(8) In a school assembly there are 350 students, all freshmen and sophomores. There are 90 more freshmen than sophomores. Three-fifths of the sophomores are girls. How many sophomore boys are there in the assembly? _____

(9) If set A is the set of all prime numbers and set B is the set of all multiples of 7, how many terms are in the intersection of A and B? _____

(10) If $a < b < 0 < c$, which of the following must be true? _____

      I. $c^b < c$

      II. $b - a > o$

      III. $b^2 > a$

(1) If the remainder when $j + 3$ is divided by 5 is 3 we know that $\dfrac{(j+3)}{5} = n + \dfrac{3}{5}$, where n is some integer. Therefore, $\dfrac{(j+3)-3}{5} = n \rightarrow \dfrac{j}{5} = n \rightarrow j = 5n$. Let's allow $n = 1$. So, $j = 5(1) \rightarrow j = 5$ and $j^2 = 25$. In fact, j can equal any square of a multiple of 5.

(2) According to the addition shown, we know that $A + B = 1C$ and $1 + A + B = AA$. So, $A + B = AA - 1$. Using substitution, we know, $AA - 1 = 1C$. Only on two-digit number whose digits are identical will produce, when 1 is subtracted from it, a result that is a two digit number beginning with 1. So, $A = 1$ and $A - 1 = C \rightarrow C = 0$. (**_Note:_** You may adopt an alternate approach. Conceptually, we know that AA and BB are two digit numbers, so their sum can be in the range of 100 to 198. Therefore, A, the hundreds-digits of the sum MUST be 1. So, $B = 9$ because 99 is the only number with the same digits in the ones and tens place that can be added to AA=11 to equal a three-digit number.

Therefore, A=1, B=9, C=0)

(3) First we want to find Cory's percent increase *this* year by solving for what percent of $60,000 is $85,000? We do the math: Allow Cory's increase to be $x\%$, and solve for **x**:

$60000(\%_{100}) = 85000 \rightarrow x = 141\frac{2}{3}$. Now, we want next year's income if his increase was the same, **x%**.

What is $141\frac{2}{3}\%$ of $85,000? $I_{nextyear} = \dfrac{x}{100}(85,000) = \dfrac{141\frac{2}{3}}{100}(85,000) = \$120,416.67$.

(4) The following sequence, $8, 3, -1, 2, 4, 5, 8, 3, -1, 2, 4, 5$ includes a repetition of the same 6 terms: $a_1 \ldots a_6$. So, the 246th term, a_{246}, will be the same as a_6 which is 5. Following the sequence, $a_{247} = 8$, $a_{248} = 3$ and $a_{249} = -1$.

(5) If James buys all cars numbered 18 to 33 inclusive he buys: $(33-18)+1$ cars (because we must include the 18th car in the count. Therefore, he buys 16 cars- $\%_8 th$ of them are grey: Let's do the math: $\frac{5}{8}(16) = 10$. Therefore, $16 - 10 = 6$ cars are magenta.

(6) Given that n is any integer, we know that $2n$ is an even integer. So, let's allow $2n + 1$ be our first odd number in our set of consecutive odd numbers. Therefore, our four consecutive odd integers are: $2n + 1$, $2n + 3$, $2n + 5$, and $2n + 7$. If their average is 32 we know: $\frac{2n + 1 + 2n + 3 + 2n + 5 + 2n + 7}{4} = 32$. Now, let's solve for n:

$\frac{8n + 16}{4} = 32 \rightarrow 8n + 16 = 128 \rightarrow 8n = 112 \rightarrow n = 14$. So the least of these integers is: $2(14) + 1 = 29$ and the greatest of these integers is: $2(14) + 7 = 35$. Now, their product$= (29)(35) = 1015$

(7) For this question, we need to know about our prime numbers from 1 to 100! Know them! Remember− a prime number is a positive integer that is divisible by only 1 and itself! We know that the greatest prime number less than 93 is **89** and the least prime number greater than 50 is **53**! Therefore, their sum is: $53 + 89 = 142$

(8) Let's allow F represent the number of freshman at the assembly and S the number of sophomores. Since there are 350 students at the assembly and they are all freshmen and sophomores, we can say: $S + F = 350$. Also, there 90 more freshman than sophomores, so: $F = 90 + S$. We'll now substitute $90 + S$ in for F in the first equation and solve for S: $S + 90 + S = 350 \rightarrow 2S = 260 \rightarrow S = 130$. Now we know that there are 130 sophomores- but three-fifths of them are girls, so let's do the math: What is $\frac{3}{5}ths$ of 130?: $S_{girls} = \frac{3}{5}(130) = 78$. Now we calculate the amount of sophomore **boys** there are at the assembly:

$S_{boys} = 130 - S_{girls} \rightarrow S_{boys} = 130 - 78 = 52$.

(9) Don't be tricked by this question! We know that the set of all prime numbers, A, is a set of all natural numbers that are divisible by only 1 and itself! So, 7 is a prime number. 7 is also in set B because it is a multiple of 7. Therefore, the only number in the intersection of set A, all prime numbers, and set B, all multiples of 7, **is** 7! We would write: $A \cap B = \{7\}$. In other words, there is only *one* term in the intersection of A and B.

(10) Since $a < b < 0 < c$ we know that both a and b are negative numbers (<0) while c is a positive number. Let's start with "I. $c^b < c$". We really can't be sure that $c^b < c$ is true because when c is a positive fraction the inequality does not hold true! Try a value: allow $c = \frac{2}{5}$, and because $b < 0$ we know c^b would equal $(\frac{2}{5})^{|b|}$ which is greater than c. (*note- you may want to review your exponent rules) Now, let's take a look at "II. $b - a > 0$". Since $a < b < 0$, we know that: $a < b < 0 \rightarrow 0 < b - a < -a$, by subtracting a throughout the inequality. So, "II. $b - a > 0$" must be true! Let's take a look at "III. $b^2 > a$": First, we know that any number squared MUST be positive, so: $b^2 > 0$, but $a < b < 0 < c$ tells us that $a < 0$ so, "III. $b^2 > a$" must be true!

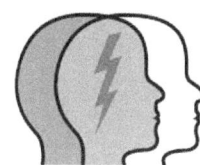

DATA, SETS & PROPORTIONALITY

Directions: Answer the following questions by showing all your work.
Note that figures are **not** drawn to scale.
Be sure to use correct units in your answers.

(1) What is the median of the set of all positive multiples of 7 less than 100?_____

(2) Given the following set: $A = \left\{8, -13, 7, \frac{2}{5}, -1, -3, 4, -1, 2, -1\right\}$, what is the (a) mean_____

(b) median_____ (c) mode_____ (d) range_____

(3) James has the following test grades so far: 73, 80, 92, and 82. What does he need to get on his last exam to finish with a 75 average? _____

(4) What is the mean of j, $j+8$ and $j-16$? _____

(5) Set M is the set of all prime numbers between -8 and 68, inclusive. Set N is the set of all the factors of 84. How many terms are in the intersection of sets M and N? _____ How many are in the union? _____

(6) The number of Chihuahuas is inversely related to the number of hyenas. When there are 16 Chihuahuas, there are 20 hyenas. How many Chihuahuas are there when there are 80 hyenas? _____

(7) If $j^8 = k^4$ find the value of j^{11}. _____

165

(1) We begin by creating set A, the numerically ordered set of all multiples of 7 less than 100:

$A = \{7,14,21,28,35,42,49,56,63,70,77,84,91,98\}$. The **median** of a set of numbers is defined as the middle term. In other words, when the set is numerically ordered, the median is the term that is exactly in the middle of the set when there are an odd number of terms) or the average of the two middle terms when there are an even number of terms. Set A has 14 elements- therefore, the median is the average between the 7th (49) and 8th (56) terms. So, $A_{median} = \dfrac{49+56}{2} = \dfrac{105}{2} \rightarrow A_{median} = 52.5$

(2) The **mean** of set $A = \{8,-13,7,\frac{2}{5},-1,-3,4,-1,2,-1\}$ is the average of its terms:

$A_{mean} = \dfrac{8 + (-13) + 7 + \frac{2}{5} + (-1) + (-3) + 4 + (-1) + 2 + (-1)}{10} = \dfrac{12}{50} = \dfrac{6}{25}$. To find set A's median, we have to place its terms in numerical order: $A = \{-13,-3-1,-1,-1,\frac{2}{5},2,4,7,8\}$. As stated in question #1, the **median** is the set's 'middle term'. There are an even number of terms in the set, so we find the average between the middle two terms. In this case: $A_{median} = \dfrac{-1 + \frac{2}{5}}{2} = -\frac{3}{10}$ (5th and 6th terms are -1 and $\frac{2}{5}$). The **mode** of a set of terms is the value in the set that occurs most often. So, $A_{mode} = -1$. The **range** of a set is the difference between the highest and the lowest values of the set. Therefore, $A_{range} = 8 - (-13) = 21$

(3) To find the average of five test scores is the following: $A_{5 tests} = \dfrac{T_1 + T_2 + T_3 + T_4 + T_5}{5}$ To find what James has to get on his fifth test in order to have an average of 75, we plug in the appropriate given values and solve for T_5: $75 = \dfrac{73 + 80 + 92 + 82 + T_5}{5} \rightarrow 375 = 327 + T_5 \rightarrow T_5 = 48$

(4) The **mean** or the average of j, $j+8$ and $j-16$ is: $\dfrac{j + j + 8 + j - 16}{3} = \dfrac{3j - 8}{3} = j - \frac{8}{3}$

(5) Remember your prime numbers from 1 to 100! Let's start with set M is the set of all prime numbers between -8 and 68 inclusive: $M = \{2,3,5,7,11,13,17,19,23,29,31,37,41,43,47,53,59,61,67\}$. Now, set N is the set of all the factors of 84: $N = \{1,2,3,4,6,7,12,14,21,28,42,84\}$. There are 3 terms in M and N's intersection: $M \cap N = \{2,3,7\}$. And, there are 28 terms in M and N's union:

$M \cup N = \{1,2,3,4,5,6,7,11,12,13,14,17,19,21,23,28,29,31,37,41,42,43,47,53,59,61,67,84\}$

(6) We are given that the number of Chihuahuas is inversely related to the number of hyenas: when there are 16 Chihuahuas, there are 20 hyenas. To find how many Chihuahuas there are we set up the equation: (Allow x= number of Chihuahuas). So, $16(20) = x(80) \rightarrow x = 4$.

(7) We are given that $j^8 = k^4$. Let's raise both sides of our equation to the $\frac{1}{8}$th power to find the value of j : $(j^8)^{\frac{1}{8}} = (k^4)^{\frac{1}{8}} \rightarrow j^1 = k^{\frac{1}{2}} \rightarrow j = \sqrt{k}$. Now, raise both sides of our equation to the $11th$ power to find the value of j^{11}: $(j)^{11} = (k^{\frac{1}{2}})^{11} \rightarrow j^{11} = k^{5.5}$.

Directions: Answer the following questions by showing all your work.
Note that figures are **not** drawn to scale.
Be sure to use correct units in your answers.

(1) The hypotenuse of an isosceles right triangle is 8. What is the triangle's area? _____

(2) The lengths of the sides of a triangle are 11, 17, and c. What is the sum of the least and greatest possible integer values of c? _____

(3) The central $\angle$ of a circle is 120°. If the radius of this circle is 6, what is the length of the arc that is created by the central $\angle$? _____

(4) Circle J is inscribed in a square. $\overline{ON} = 16$ in. What is the area of the shaded region? _____

(5) What is the median of the prime numbers between -7 and 17, inclusive? _____

(6) Ken has a 73, 81, 84 and 66 on his mathematics exams. What does he need on his last exam to finish the semester with an average of 75? _____

(7) The sum of three consecutive even numbers is 78. What is the average of the greater two numbers? _____

(1) If the hypotenuse of an isosceles triangle = 8cm, we are dealing with an isosceles right triangle. This kind of triangle is frequently referred to as a 45-45-90 triangle. The ratio of the sides of an *isosceles right triangle* is

$x : x : x\sqrt{2}$. In this case, $8 = x\sqrt{2} \rightarrow x = \dfrac{8}{\sqrt{2}}$. The area of a triangle is given by the formula:

$A_\triangle = \frac{1}{2}(base)(height)$. For this isosceles right triangle, both the base and the height are the same. So,

$$A_\triangle = \frac{1}{2}(x)^2 \rightarrow A_\triangle = \frac{1}{2}\left(\frac{8}{\sqrt{2}}\right)^2 \rightarrow A_\triangle = 16$$

(2) An important rule to know about the lengths of the sides of a triangle: *the sum of the lengths of any two sides will be greater than the length of the third side*. Therefore, in this triangle with side lengths of 11, 17, and c, we know that: $17 + 11 > c \rightarrow c < 28$. So, the greatest possible integer of c is 27. We also know that:

$11 + c > 17 \rightarrow c > 6$. So, the least possible integer of c is 7. So the sum, $c_{least} + c_{greatest} = 7 + 27 = 34$

(3) Draw circle, L with a central $\angle$ of $120°$. Label the arc opposite this angle, $\overparen{AP}$

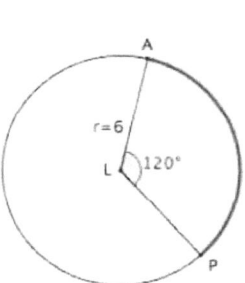

Before finding the length of $\overparen{AP}$, we must first find the circumference of the entire circle. The Circumference of a circle is given by the formula: $C_O = 2\pi r$. We find the circumference for circle L: $C_{OL} = 12\pi$. Now, we need to see what **portion** of the entire circumference is the smaller arc, $\overparen{AB}$?. Let's set up a proportion:

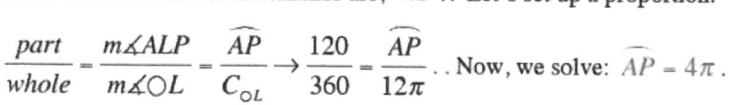

$$\frac{part}{whole} = \frac{m\angle ALP}{m\angle OL} = \frac{\overparen{AP}}{C_{OL}} \rightarrow \frac{120}{360} = \frac{\overparen{AP}}{12\pi} \quad .. \text{Now, we solve: } \overparen{AP} = 4\pi.$$

169

(4) Since $\overline{ON} = 16$, the diagonal of the square, we can find the value of one of the sides of the square. Because the diagonals of a square bisect the corner angles, we have an isosceles right triangle with hypotenuse of 16. Using the ratios of the sides of an isosceles right triangle $s : s : s\sqrt{2}$. we can solve for s: $s\sqrt{2} = 16 \rightarrow s = \frac{16}{\sqrt{2}}$.

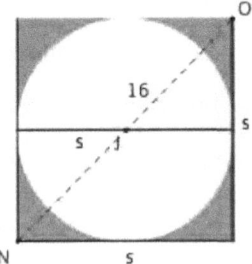

The area of a square is given by: $A_{\square} = s^2$, so $A_{\square} = \left(\frac{16}{\sqrt{2}}\right)^2 = 128$. Now, the area of this circle is given by: $A_{\odot J} = \pi r^2$. As you see in the diagram, length s is also the same as the diameter, d of circle J. And, $r = \frac{1}{2}d$. Therefore, $r = \frac{8}{\sqrt{2}}$. So, $A_{\odot J} = \left(\frac{8}{\sqrt{2}}\right)^2 \pi = 32\pi$. Now,

$A_{shadedregion} = A_{\square} - A_{\odot J} = 128 - 32\pi = 27.469$

(5) Don't be fooled by this question! You think that -7 is a prime number? Wrong! *A prime number is a positive integer greater than one that is only divisible by one and itself.* Therefore, the prime number between -7 and 17 inclusive are: $\{2, 3, 5, 7, 11, 13, 17\}$. The median of a set of numbers (in order) is the "middle number". Therefore, the median of these numbers is 7.

(6) To find the average of five test scores we'll use the following formula: $T_{Average} = \frac{T_1 + T_2 + T_3 + T_4 + T_5}{5}$. We know Ken has a 73, 81, 84 and 66 on his tests. To find what he has to score on T_5 to receive a 75 average, plug your values into the given equation and solve for T_5. $75 = \frac{73 + 81 + 84 + 66 + T_5}{5} \rightarrow 375 = 304 + T_5 \rightarrow T_5 = 71$.

(7) Let's allow $2x$, $2x + 2$, and $2x + 4$, ($x \in \mathbb{N}$), be our three consecutive even numbers. We know that $2x + 2x + 2 + 2x + 4 = 78$. To find the value of each of the greater two numbers, we want to first for x: $2x + 2x + 2 + 2x + 4 = 78 \rightarrow 6x = 72 \rightarrow x = 12$. Therefore, $2x + 2 = 26$ and $2x + 4 = 28$. So the average of these two numbers is: $A_{greatertwo} = \frac{28 + 26}{2} = 27$

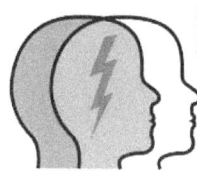

Directions: Answer the following questions by showing all your work.

Note that figures are **not** drawn to scale.

Be sure to use correct units in your answers.

(1) If $\sin A = \dfrac{a}{b}$ what is $\tan A$? _____

(2) If 2M -- K = 14 and $3K^2 - 5 = X$, find X in terms of M. _____

(3) How many diagonals are in a dodecagon? _____

(4) What is the area of this parallelogram? _____

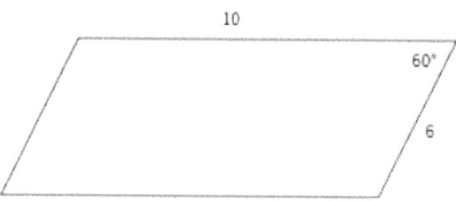

(5) Solve: 3 (x-5) + 9x ≤ 14 − 6 (5 -2x). _____

(6) What is the GCF of $18A^2B^9$ and $24AB^5C^2$? _____

(7) Simplify: $\dfrac{19x^{-2}y^4z}{57x^{-3}y^{-2}z^2}$

_____

(8) What is the area of this trapezoid? _____

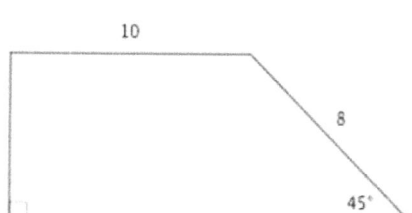

(1) The sine of an angle in a right triangle equals the ratio of the opposite side and the hypotenuse. Therefore, $\sin A = \dfrac{opp}{hyp} = \dfrac{a}{b}$. The tangent of this same angle, $\angle A$ equals the ratio of the angle's opposite side to its adjacent side. To find the value of side c (adjacent) in this triangle use **Pythagorean theorem:** $b^2 = a^2 + c^2 \rightarrow c = \sqrt{b^2 - a^2}$. Therefore, $\tan A = \dfrac{opp}{adj} \rightarrow \tan A = \dfrac{a}{\sqrt{b^2 - a^2}}$.

(2) First, let's solve $2M - K = 14$ for K: $2M - K = 14 \rightarrow K = 2M - 14$. Now, we can use substitution in the second equation: $3(2M - 14)^2 - 5 = X$ or $X = 12M^2 - 168M + 583$.

(3) The number of diagonals in a polygon with n sides is given by the formula: $D = \left(\dfrac{n}{2}\right)(n - 3)$ where D represents the number of diagonals. Since a dodecagon has 12 sides, $D_{dodec} = \left(\dfrac{12}{2}\right)(12 - 3) \rightarrow D_{dodec} = 54$.

(4) The area of a parallelogram is given by the formula: $A_{parallel} = (length)(height)$. We are given the length, l of parallelogram ABCD ($\square BADC$) to be 10. Now, we have to find the height. We know that the height can be found by first drawing line that represents the altitude of

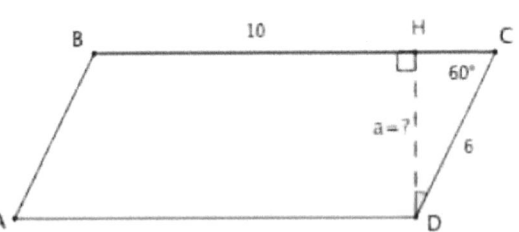

the parallelogram- that is, line $\overline{DH}$. Notice how the triangle formed, $\triangle CHD$ is a 30-60-90 triangle with side length ratio of $x : x\sqrt{3} : 2x$. We know that $6 = 2x \rightarrow x = 3$. Therefore, $\overline{DH} = x\sqrt{3} = 3\sqrt{3}$ (*how do we know this? It is across from the $60°$ angle). Therefore, $A_{\square BADC} = (10)(3\sqrt{3}) = 30\sqrt{3}$.

(5) Solve: $3(x - 5) + 9x \leq 14 - 6(5 - 2x) \rightarrow -15 + \cancel{12}x \leq -16 + \cancel{12}x \rightarrow -15 \cancel{\leq} -16$. Because there is no x value for which this inequality is true- our answer is *no solution.*

(6) To find the GCF (**Greatest Common Factor**) of $18A^2B^9$ and $24AB^5C^2$ you want to find the biggest number that divides into both terms. In this case it is: $6AB^5$.

WELL DONE!!

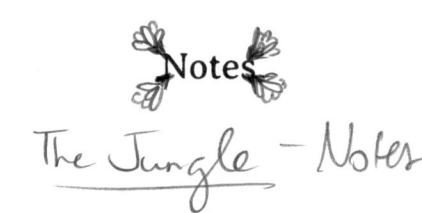

The Jungle - Notes

Chapter 1

The Jungle → condemn the evils of Capitalism

Wedding party → full of Lithuanian immigrants who came for American dream

<u>Jurgis</u> (the groom) → hopeful, naive and believes hard work will pay off

• Details of wedding - foreshadow problems later in the in the novel to be exposed

<u>Antanos</u>: six months in Chicago, grueling job in pickle room

Jurgis + Ona → rely on money from tradition to
   ↓
   bride
cover the wedding

   Guests: too poor, overworked, depressed

<u>Barmans debt</u> → Barman is well connected in the district, has power, his debt/bill is unchallangeable

   ⟹ Barmans debt hints at role of the Democratic-run political machine

<u>Couple</u> → has more expenses since America

   • Policemen in Lithuania
   • Border agents in NY
   • Agreements, to divide market for rent/collusion

• No matter how hard you work → no output

· Lithuanians → moving away from traditions, moving toward
fraud + crime

New modern ways : getting something for nothing

· Sacrifice of culture → everyone in harm

〰〰〰〰〰

Notes

Notes